T0198079

Get the eBooks FREE!

(PDF, ePub, Kindle, and liveBook all included)

We believe that once you buy a book from us, you should be able to read it in any format we have available. To get electronic versions of this book at no additional cost to you, purchase and then register this book at the Manning website.

Go to https://www.manning.com/freebook and follow the instructions to complete your pBook registration.

That's it!
Thanks from Manning!

Introducing Data Science

BIG DATA, MACHINE LEARNING, AND MORE, USING PYTHON TOOLS

DAVY CIELEN
ARNO D. B. MEYSMAN
MOHAMED ALI

MANNING
SHELTER ISLAND

For online information and ordering of this and other Manning books, please visit www.manning.com. The publisher offers discounts on this book when ordered in quantity. For more information, please contact

> Special Sales Department
> Manning Publications Co.
> 20 Baldwin Road
> PO Box 761
> Shelter Island, NY 11964
> Email: orders@manning.com

Manning Publications Co.
20 Baldwin Road
PO Box 761
Shelter Island, NY 11964

Development editor:	Dan Maharry
Technical development editors:	Michael Roberts, Jonathan Thoms
Copyeditor:	Katie Petito
Proofreader:	Alyson Brener
Technical proofreader:	Ravishankar Rajagopalan
Typesetter:	Dennis Dalinnik
Cover designer:	Marija Tudor

ISBN: 9781633430037
Printed in the United States of America

brief contents

contents

2 The data science process   22

3 Machine learning   57

9 Data visualization to the end user 253

preface

It's in all of us. Data science is what makes us humans what we are today. No, not the computer-driven data science this book will introduce you to, but the ability of our brains to see connections, draw conclusions from facts, and learn from our past experiences. More so than any other species on the planet, we depend on our brains for survival; we went all-in on these features to earn our place in nature. That strategy has worked out for us so far, and we're unlikely to change it in the near future.

But our brains can only take us so far when it comes to raw computing. Our biology can't keep up with the amounts of data we can capture now and with the extent of our curiosity. So we turn to machines to do part of the work for us: to recognize patterns, create connections, and supply us with answers to our numerous questions.

The quest for knowledge is in our genes. Relying on computers to do part of the job for us is not—but it is our destiny.

acknowledgments

A big thank you to all the people of Manning involved in the process of making this book for guiding us all the way through.

Our thanks also go to Ravishankar Rajagopalan for giving the manuscript a full technical proofread, and to Jonathan Thoms and Michael Roberts for their expert comments. There were many other reviewers who provided invaluable feedback throughout the process: Alvin Raj, Arthur Zubarev, Bill Martschenko, Craig Smith, Filip Pravica, Hamideh Iraj, Heather Campbell, Hector Cuesta, Ian Stirk, Jeff Smith, Joel Kotarski, Jonathan Sharley, Jörn Dinkla, Marius Butuc, Matt R. Cole, Matthew Heck, Meredith Godar, Rob Agle, Scott Chaussee, and Steve Rogers.

First and foremost I want to thank my wife Filipa for being my inspiration and motivation to beat all difficulties and for always standing beside me throughout my career and the writing of this book. She has provided me the necessary time to pursue my goals and ambition, and shouldered all the burdens of taking care of our little daughter in my absence. I dedicate this book to her and really appreciate all the sacrifices she has made in order to build and maintain our little family.

I also want to thank my daughter Eva, and my son to be born, who give me a great sense of joy and keep me smiling. They are the best gifts that God ever gave to my life and also the best children a dad could hope for: fun, loving, and always a joy to be with.

A special thank you goes to my parents for their support over the years. Without the endless love and encouragement from my family, I would not have been able to finish this book and continue the journey of achieving my goals in life.

I'd really like to thank all my coworkers in my company, especially Mo and Arno, for all the adventures we have been through together. Mo and Arno have provided me excellent support and advice. I appreciate all of their time and effort in making this book complete. They are great people, and without them, this book may not have been written.

Finally, a sincere thank you to my friends who support me and understand that I do not have much time but I still count on the love and support they have given me throughout my career and the development of this book.

Davy Cielen

I would like to give thanks to my family and friends who have supported me all the way through the process of writing this book. It has not always been easy to stay at home writing, while I could be out discovering new things. I want to give very special thanks to my parents, my brother Jago, and my girlfriend Delphine for always being there for me, regardless of what crazy plans I come up with and execute.

I would also like to thank my godmother, and my godfather whose current struggle with cancer puts everything in life into perspective again.

Thanks also go to my friends for buying me beer to distract me from my work and to Delphine's parents, her brother Karel, and his soon-to-be wife Tess for their hospitality (and for stuffing me with good food).

All of them have made a great contribution to a wonderful life so far.

Last but not least, I would like to thank my coauthor Mo, my ERC-homie, and my coauthor Davy for their insightful contributions to this book. I share the ups and downs of being an entrepreneur and data scientist with both of them on a daily basis. It has been a great trip so far. Let's hope there are many more days to come.

Arno D. B. Meysman

First and foremost, I would like to thank my fiancée Muhuba for her love, understanding, caring, and patience. Finally, I owe much to Davy and Arno for having fun and for making an entrepreneurial dream come true. Their unfailing dedication has been a vital resource for the realization of this book.

Mohamed Ali

about this book

I can only show you the door. You're the one that has to walk through it.

Morpheus, *The Matrix*

Welcome to the book! When reading the table of contents, you probably noticed the diversity of the topics we're about to cover. The goal of *Introducing Data Science* is to provide you with a little bit of everything—enough to get you started. Data science is a very wide field, so wide indeed that a book ten times the size of this one wouldn't be able to cover it all. For each chapter, we picked a different aspect we find interesting. Some hard decisions had to be made to keep this book from collapsing your bookshelf!

We hope it serves as an entry point—your doorway into the exciting world of data science.

Roadmap

Chapters 1 and 2 offer the general theoretical background and framework necessary to understand the rest of this book:

- Chapter 1 is an introduction to data science and big data, ending with a practical example of Hadoop.
- Chapter 2 is all about the data science process, covering the steps present in almost every data science project.

In chapters 3 through 5, we apply machine learning on increasingly large data sets:

- Chapter 3 keeps it small. The data still fits easily into an average computer's memory.
- Chapter 4 increases the challenge by looking at "large data." This data fits on your machine, but fitting it into RAM is hard, making it a challenge to process without a computing cluster.
- Chapter 5 finally looks at big data. For this we can't get around working with multiple computers.

Chapters 6 through 9 touch on several interesting subjects in data science in a more-or-less independent matter:

- Chapter 6 looks at NoSQL and how it differs from the relational databases.
- Chapter 7 applies data science to streaming data. Here the main problem is not size, but rather the speed at which data is generated and old data becomes obsolete.
- Chapter 8 is all about text mining. Not all data starts off as numbers. Text mining and text analytics become important when the data is in textual formats such as emails, blogs, websites, and so on.
- Chapter 9 focuses on the last part of the data science process—data visualization and prototype application building—by introducing a few useful HTML5 tools.

Appendixes A–D cover the installation and setup of the Elasticsearch, Neo4j, and MySQL databases described in the chapters and of Anaconda, a Python code package that's especially useful for data science.

Whom this book is for

This book is an introduction to the field of data science. Seasoned data scientists will see that we only scratch the surface of some topics. For our other readers, there are some prerequisites for you to fully enjoy the book. A minimal understanding of SQL, Python, HTML5, and statistics or machine learning is recommended before you dive into the practical examples.

Code conventions and downloads

We opted to use the Python script for the practical examples in this book. Over the past decade, Python has developed into a much respected and widely used data science language.

The code itself is presented in a `fixed-width font like this` to separate it from ordinary text. Code annotations accompany many of the listings, highlighting important concepts.

The book contains many code examples, most of which are available in the online code base, which can be found at the book's website, https://www.manning.com/books/introducing-data-science.

about the authors

DAVY CIELEN is an experienced entrepreneur, book author, and professor. He is the co-owner with Arno and Mo of Optimately and Maiton, two data science companies based in Belgium and the UK, respectively, and co-owner of a third data science company based in Somaliland. The main focus of these companies is on strategic big data science, and they are occasionally consulted by many large companies. Davy is an adjunct professor at the IESEG School of Management in Lille, France, where he is involved in teaching and research in the field of big data science.

ARNO MEYSMAN is a driven entrepreneur and data scientist. He is the co-owner with Davy and Mo of Optimately and Maiton, two data science companies based in Belgium and the UK, respectively, and co-owner of a third data science company based in Somaliland. The main focus of these companies is on strategic big data science, and they are occasionally consulted by many large companies. Arno is a data scientist with a wide spectrum of interests, ranging from medical analysis to retail to game analytics. He believes insights from data combined with some imagination can go a long way toward helping us to improve this world.

MOHAMED ALI is an entrepreneur and a data science consultant. Together with Davy and Arno, he is the co-owner of Optimately and Maiton, two data science companies based in Belgium and the UK, respectively. His passion lies in two areas, data science and sustainable projects, the latter being materialized through the creation of a third company based in Somaliland.

Author Online

The purchase of *Introducing Data Science* includes free access to a private web forum run by Manning Publications where you can make comments about the book, ask technical questions, and receive help from the lead author and from other users. To access the forum and subscribe to it, point your web browser to https://www.manning .com/books/introducing-data-science. This page provides information on how to get on the forum once you are registered, what kind of help is available, and the rules of conduct on the forum.

Manning's commitment to our readers is to provide a venue where a meaningful dialog between individual readers and between readers and the author can take place. It is not a commitment to any specific amount of participation on the part of the author, whose contribution to AO remains voluntary (and unpaid). We suggest you try asking the author some challenging questions lest his interest stray! The Author Online forum and the archives of previous discussions will be accessible from the publisher's website as long as the book is in print.

about the cover illustration

The illustration on the cover of *Introducing Data Science* is taken from the 1805 edition of Sylvain Maréchal's four-volume compendium of regional dress customs. This book was first published in Paris in 1788, one year before the French Revolution. Each illustration is colored by hand. The caption for this illustration reads "Homme Salamanque," which means man from Salamanca, a province in western Spain, on the border with Portugal. The region is known for its wild beauty, lush forests, ancient oak trees, rugged mountains, and historic old towns and villages.

The Homme Salamanque is just one of many figures in Maréchal's colorful collection. Their diversity speaks vividly of the uniqueness and individuality of the world's towns and regions just 200 years ago. This was a time when the dress codes of two regions separated by a few dozen miles identified people uniquely as belonging to one or the other. The collection brings to life a sense of the isolation and distance of that period and of every other historic period—except our own hyperkinetic present.

Dress codes have changed since then and the diversity by region, so rich at the time, has faded away. It is now often hard to tell the inhabitant of one continent from another. Perhaps we have traded cultural diversity for a more varied personal life—certainly for a more varied and fast-paced technological life.

We at Manning celebrate the inventiveness, the initiative, and the fun of the computer business with book covers based on the rich diversity of regional life two centuries ago, brought back to life by Maréchal's pictures.

Data science in
a big data world

This chapter covers

- Defining data science and big data
- Recognizing the different types of data
- Gaining insight into the data science process
- Introducing the fields of data science and big data
- Working through examples of Hadoop

Big data is a blanket term for any collection of data sets so large or complex that it becomes difficult to process them using traditional data management techniques such as, for example, the RDBMS (relational database management systems). The widely adopted RDBMS has long been regarded as a one-size-fits-all solution, but the demands of handling big data have shown otherwise. *Data science* involves using methods to analyze massive amounts of data and extract the knowledge it contains. You can think of the relationship between big data and data science as being like the relationship between crude oil and an oil refinery. Data science and big data evolved from statistics and traditional data management but are now considered to be distinct disciplines.

1

The characteristics of big data are often referred to as the three Vs:

- *Volume*—How much data is there?
- *Variety*—How diverse are different types of data?
- *Velocity*—At what speed is new data generated?

Often these characteristics are complemented with a fourth V, veracity: How accurate is the data? These four properties make big data different from the data found in traditional data management tools. Consequently, the challenges they bring can be felt in almost every aspect: data capture, curation, storage, search, sharing, transfer, and visualization. In addition, big data calls for specialized techniques to extract the insights.

Data science is an evolutionary extension of statistics capable of dealing with the massive amounts of data produced today. It adds methods from computer science to the repertoire of statistics. In a research note from Laney and Kart, *Emerging Role of the Data Scientist and the Art of Data Science*, the authors sifted through hundreds of job descriptions for data scientist, statistician, and BI (Business Intelligence) analyst to detect the differences between those titles. The main things that set a data scientist apart from a statistician are the ability to work with big data and experience in machine learning, computing, and algorithm building. Their tools tend to differ too, with data scientist job descriptions more frequently mentioning the ability to use Hadoop, Pig, Spark, R, Python, and Java, among others. Don't worry if you feel intimidated by this list; most of these will be gradually introduced in this book, though we'll focus on Python. Python is a great language for data science because it has many data science libraries available, and it's widely supported by specialized software. For instance, almost every popular NoSQL database has a Python-specific API. Because of these features and the ability to prototype quickly with Python while keeping acceptable performance, its influence is steadily growing in the data science world.

As the amount of data continues to grow and the need to leverage it becomes more important, every data scientist will come across big data projects throughout their career.

1.1 Benefits and uses of data science and big data

Data science and big data are used almost everywhere in both commercial and noncommercial settings. The number of use cases is vast, and the examples we'll provide throughout this book only scratch the surface of the possibilities.

Commercial companies in almost every industry use data science and big data to gain insights into their customers, processes, staff, completion, and products. Many companies use data science to offer customers a better user experience, as well as to cross-sell, up-sell, and personalize their offerings. A good example of this is Google AdSense, which collects data from internet users so relevant commercial messages can be matched to the person browsing the internet. MaxPoint (http://maxpoint.com/us)

is another example of real-time personalized advertising. Human resource professionals use people analytics and text mining to screen candidates, monitor the mood of employees, and study informal networks among coworkers. People analytics is the central theme in the book *Moneyball: The Art of Winning an Unfair Game.* In the book (and movie) we saw that the traditional scouting process for American baseball was random, and replacing it with correlated signals changed everything. Relying on statistics allowed them to hire the right players and pit them against the opponents where they would have the biggest advantage. Financial institutions use data science to predict stock markets, determine the risk of lending money, and learn how to attract new clients for their services. At the time of writing this book, at least 50% of trades worldwide are performed automatically by machines based on algorithms developed by *quants*, as data scientists who work on trading algorithms are often called, with the help of big data and data science techniques.

Governmental organizations are also aware of data's value. Many governmental organizations not only rely on internal data scientists to discover valuable information, but also share their data with the public. You can use this data to gain insights or build data-driven applications. *Data.gov* is but one example; it's the home of the US Government's open data. A data scientist in a governmental organization gets to work on diverse projects such as detecting fraud and other criminal activity or optimizing project funding. A well-known example was provided by Edward Snowden, who leaked internal documents of the American National Security Agency and the British Government Communications Headquarters that show clearly how they used data science and big data to monitor millions of individuals. Those organizations collected 5 billion data records from widespread applications such as Google Maps, Angry Birds, email, and text messages, among many other data sources. Then they applied data science techniques to distill information.

Nongovernmental organizations (NGOs) are also no strangers to using data. They use it to raise money and defend their causes. The World Wildlife Fund (WWF), for instance, employs data scientists to increase the effectiveness of their fundraising efforts. Many data scientists devote part of their time to helping NGOs, because NGOs often lack the resources to collect data and employ data scientists. DataKind is one such data scientist group that devotes its time to the benefit of mankind.

Universities use data science in their research but also to enhance the study experience of their students. The rise of massive open online courses (MOOC) produces a lot of data, which allows universities to study how this type of learning can complement traditional classes. MOOCs are an invaluable asset if you want to become a data scientist and big data professional, so definitely look at a few of the better-known ones: Coursera, Udacity, and edX. The big data and data science landscape changes quickly, and MOOCs allow you to stay up to date by following courses from top universities. If you aren't acquainted with them yet, take time to do so now; you'll come to love them as we have.

1.2   *Facets of data*

In data science and big data you'll come across many different types of data, and each of them tends to require different tools and techniques. The main categories of data are these:

- Structured
- Unstructured
- Natural language
- Machine-generated
- Graph-based
- Audio, video, and images
- Streaming

Let's explore all these interesting data types.

1.2.1   *Structured data*

Structured data is data that depends on a data model and resides in a fixed field within a record. As such, it's often easy to store structured data in tables within databases or Excel files (figure 1.1). SQL, or Structured Query Language, is the preferred way to manage and query data that resides in databases. You may also come across structured data that might give you a hard time storing it in a traditional relational database. Hierarchical data such as a family tree is one such example.

The world isn't made up of structured data, though; it's imposed upon it by humans and machines. More often, data comes unstructured.

1	Indicator ID	Dimension List	Timeframe	Numeric Value	Missing Value Flag	Confidence Inte
2	214390830	Total (Age-adjusted)	2008	74.6%		73.8%
3	214390833	Aged 18-44 years	2008	59.4%		58.0%
4	214390831	Aged 18-24 years	2008	37.4%		34.6%
5	214390832	Aged 25-44 years	2008	66.9%		65.5%
6	214390836	Aged 45-64 years	2008	88.6%		87.7%
7	214390834	Aged 45-54 years	2008	86.3%		85.1%
8	214390835	Aged 55-64 years	2008	91.5%		90.4%
9	214390840	Aged 65 years and over	2008	94.6%		93.8%
10	214390837	Aged 65-74 years	2008	93.6%		92.4%
11	214390838	Aged 75-84 years	2008	95.6%		94.4%
12	214390839	Aged 85 years and over	2008	96.0%		94.0%
13	214390841	Male (Age-adjusted)	2008	72.2%		71.1%
14	214390842	Female (Age-adjusted)	2008	76.8%		75.9%
15	214390843	White only (Age-adjusted)	2008	73.8%		72.9%
16	214390844	Black or African American only (Age-adjusted)	2008	77.0%		75.0%
17	214390845	American Indian or Alaska Native only (Age-adjusted)	2008	66.5%		57.1%
18	214390846	Asian only (Age-adjusted)	2008	80.5%		77.7%
19	214390847	Native Hawaiian or Other Pacific Islander only (Age-adjusted)	2008	DSU		
20	214390848	2 or more races (Age-adjusted)	2008	75.6%		69.6%

Figure 1.1   An Excel table is an example of structured data.

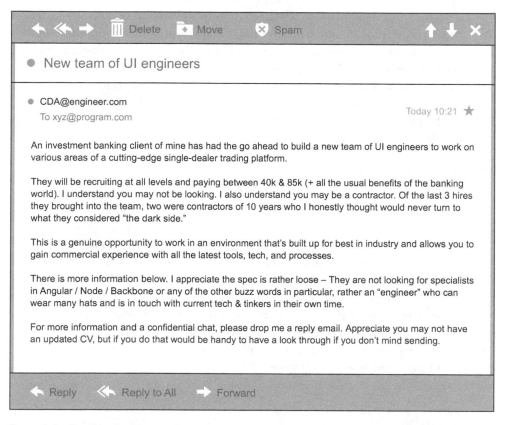

Figure 1.2   Email is simultaneously an example of unstructured data and natural language data.

1.2.2   *Unstructured data*

Unstructured data is data that isn't easy to fit into a data model because the content is context-specific or varying. One example of unstructured data is your regular email (figure 1.2). Although email contains structured elements such as the sender, title, and body text, it's a challenge to find the number of people who have written an email complaint about a specific employee because so many ways exist to refer to a person, for example. The thousands of different languages and dialects out there further complicate this.

A human-written email, as shown in figure 1.2, is also a perfect example of natural language data.

1.2.3   *Natural language*

Natural language is a special type of unstructured data; it's challenging to process because it requires knowledge of specific data science techniques and linguistics.

The natural language processing community has had success in entity recognition, topic recognition, summarization, text completion, and sentiment analysis, but models trained in one domain don't generalize well to other domains. Even state-of-the-art techniques aren't able to decipher the meaning of every piece of text. This shouldn't be a surprise though: humans struggle with natural language as well. It's ambiguous by nature. The concept of meaning itself is questionable here. Have two people listen to the same conversation. Will they get the same meaning? The meaning of the same words can vary when coming from someone upset or joyous.

1.2.4 *Machine-generated data*

Machine-generated data is information that's automatically created by a computer, process, application, or other machine without human intervention. Machine-generated data is becoming a major data resource and will continue to do so. Wikibon has forecast that the market value of the *industrial Internet* (a term coined by Frost & Sullivan to refer to the integration of complex physical machinery with networked sensors and software) will be approximately $540 billion in 2020. IDC (International Data Corporation) has estimated there will be 26 times more connected things than people in 2020. This network is commonly referred to as *the internet of things.*

The analysis of machine data relies on highly scalable tools, due to its high volume and speed. Examples of machine data are web server logs, call detail records, network event logs, and telemetry (figure 1.3).

```
CSIPERF:TXCOMMIT;313236
2014-11-28 11:36:13, Info          CSI    00000153 Creating NT transaction (seq
69), objectname [6]"(null)"
2014-11-28 11:36:13, Info          CSI    00000154 Created NT transaction (seq 69)
result 0x00000000, handle @0x4e54
2014-11-28 11:36:13, Info          CSI    00000155@2014/11/28:10:36:13.471
Beginning NT transaction commit...
2014-11-28 11:36:13, Info          CSI    00000156@2014/11/28:10:36:13.705 CSI perf
trace:
CSIPERF:TXCOMMIT;273983
2014-11-28 11:36:13, Info          CSI    00000157 Creating NT transaction (seq
70), objectname [6]"(null)"
2014-11-28 11:36:13, Info          CSI    00000158 Created NT transaction (seq 70)
result 0x00000000, handle @0x4e5c
2014-11-28 11:36:13, Info          CSI    00000159@2014/11/28:10:36:13.764
Beginning NT transaction commit...
2014-11-28 11:36:14, Info          CSI    0000015a@2014/11/28:10:36:14.094 CSI perf
trace:
CSIPERF:TXCOMMIT;386259
2014-11-28 11:36:14, Info          CSI    0000015b Creating NT transaction (seq
71), objectname [6]"(null)"
2014-11-28 11:36:14, Info          CSI    0000015c Created NT transaction (seq 71)
result 0x00000000, handle @0x4e5c
2014-11-28 11:36:14, Info          CSI    0000015d@2014/11/28:10:36:14.106
Beginning NT transaction commit...
2014-11-28 11:36:14, Info          CSI    0000015e@2014/11/28:10:36:14.428 CSI perf
trace:
CSIPERF:TXCOMMIT;375581
```

Figure 1.3 **Example of machine-generated data**

The machine data shown in figure 1.3 would fit nicely in a classic table-structured database. This isn't the best approach for highly interconnected or "networked" data, where the relationships between entities have a valuable role to play.

1.2.5 *Graph-based or network data*

"Graph data" can be a confusing term because any data can be shown in a graph. "Graph" in this case points to mathematical *graph theory*. In graph theory, a graph is a mathematical structure to model pair-wise relationships between objects. Graph or network data is, in short, data that focuses on the relationship or adjacency of objects. The graph structures use nodes, edges, and properties to represent and store graphical data. Graph-based data is a natural way to represent social networks, and its structure allows you to calculate specific metrics such as the influence of a person and the shortest path between two people.

Examples of graph-based data can be found on many social media websites (figure 1.4). For instance, on LinkedIn you can see who you know at which company. Your follower list on Twitter is another example of graph-based data. The power and sophistication comes from multiple, overlapping graphs of the same nodes. For example, imagine the connecting edges here to show "friends" on Facebook. Imagine another graph with the same people which connects business colleagues via LinkedIn. Imagine a third graph based on movie interests on Netflix. Overlapping the three different-looking graphs makes more interesting questions possible.

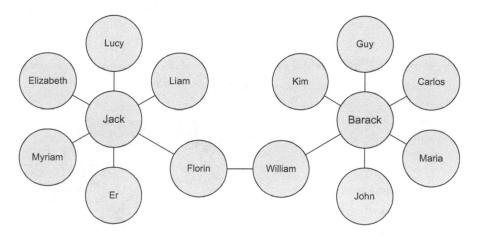

Figure 1.4 Friends in a social network are an example of graph-based data.

Graph databases are used to store graph-based data and are queried with specialized query languages such as SPARQL.

Graph data poses its challenges, but for a computer interpreting additive and image data, it can be even more difficult.

1.2.6 *Audio, image, and video*

Audio, image, and video are data types that pose specific challenges to a data scientist. Tasks that are trivial for humans, such as recognizing objects in pictures, turn out to be challenging for computers. MLBAM (Major League Baseball Advanced Media) announced in 2014 that they'll increase video capture to approximately 7 TB per game for the purpose of live, in-game analytics. High-speed cameras at stadiums will capture ball and athlete movements to calculate in real time, for example, the path taken by a defender relative to two baselines.

Recently a company called DeepMind succeeded at creating an algorithm that's capable of learning how to play video games. This algorithm takes the video screen as input and learns to interpret everything via a complex process of deep learning. It's a remarkable feat that prompted Google to buy the company for their own Artificial Intelligence (AI) development plans. The learning algorithm takes in data as it's produced by the computer game; it's streaming data.

1.2.7 *Streaming data*

While streaming data can take almost any of the previous forms, it has an extra property. The data flows into the system when an event happens instead of being loaded into a data store in a batch. Although this isn't really a different type of data, we treat it here as such because you need to adapt your process to deal with this type of information.

Examples are the "What's trending" on Twitter, live sporting or music events, and the stock market.

1.3   *The data science process*

The data science process typically consists of six steps, as you can see in the mind map in figure 1.5. We will introduce them briefly here and handle them in more detail in chapter 2.

1.3.1 *Setting the research goal*

Data science is mostly applied in the context of an organization. When the business asks you to perform a data science project, you'll first prepare a project charter. This charter contains information such as what you're going to research, how the company benefits from that, what data and resources you need, a timetable, and deliverables.

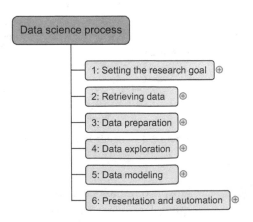

Figure 1.5 **The data science process**

Throughout this book, the data science process will be applied to bigger case studies and you'll get an idea of different possible research goals.

1.3.2 Retrieving data

The second step is to collect data. You've stated in the project charter which data you need and where you can find it. In this step you ensure that you can use the data in your program, which means checking the existence of, quality, and access to the data. Data can also be delivered by third-party companies and takes many forms ranging from Excel spreadsheets to different types of databases.

1.3.3 Data preparation

Data collection is an error-prone process; in this phase you enhance the quality of the data and prepare it for use in subsequent steps. This phase consists of three sub-phases: *data cleansing* removes false values from a data source and inconsistencies across data sources, *data integration* enriches data sources by combining information from multiple data sources, and *data transformation* ensures that the data is in a suitable format for use in your models.

1.3.4 Data exploration

Data exploration is concerned with building a deeper understanding of your data. You try to understand how variables interact with each other, the distribution of the data, and whether there are outliers. To achieve this you mainly use descriptive statistics, visual techniques, and simple modeling. This step often goes by the abbreviation EDA, for Exploratory Data Analysis.

1.3.5 Data modeling or model building

In this phase you use models, domain knowledge, and insights about the data you found in the previous steps to answer the research question. You select a technique from the fields of statistics, machine learning, operations research, and so on. Building a model is an iterative process that involves selecting the variables for the model, executing the model, and model diagnostics.

1.3.6 Presentation and automation

Finally, you present the results to your business. These results can take many forms, ranging from presentations to research reports. Sometimes you'll need to automate the execution of the process because the business will want to use the insights you gained in another project or enable an operational process to use the outcome from your model.

AN ITERATIVE PROCESS   The previous description of the data science process gives you the impression that you walk through this process in a linear way, but in reality you often have to step back and rework certain findings. For instance, you might find outliers in the data exploration phase that point to data import errors. As part of the data science process you gain incremental insights, which may lead to new questions. To prevent rework, make sure that you scope the business question clearly and thoroughly at the start.

Now that we have a better understanding of the process, let's look at the technologies.

1.4   *The big data ecosystem and data science*

Currently many big data tools and frameworks exist, and it's easy to get lost because new technologies appear rapidly. It's much easier once you realize that the big data ecosystem can be grouped into technologies that have similar goals and functionalities, which we'll discuss in this section. Data scientists use many different technologies, but not all of them; we'll dedicate a separate chapter to the most important data science technology classes. The mind map in figure 1.6 shows the components of the big data ecosystem and where the different technologies belong.

Let's look at the different groups of tools in this diagram and see what each does. We'll start with distributed file systems.

1.4.1   *Distributed file systems*

A *distributed file system* is similar to a normal file system, except that it runs on multiple servers at once. Because it's a file system, you can do almost all the same things you'd do on a normal file system. Actions such as storing, reading, and deleting files and adding security to files are at the core of every file system, including the distributed one. Distributed file systems have significant advantages:

- They can store files larger than any one computer disk.
- Files get automatically replicated across multiple servers for redundancy or parallel operations while hiding the complexity of doing so from the user.
- The system scales easily: you're no longer bound by the memory or storage restrictions of a single server.

In the past, scale was increased by moving everything to a server with more memory, storage, and a better CPU (vertical scaling). Nowadays you can add another small server (horizontal scaling). This principle makes the scaling potential virtually limitless.

The best-known distributed file system at this moment is the *Hadoop File System (HDFS)*. It is an open source implementation of the Google File System. In this book we focus on the Hadoop File System because it is the most common one in use. However, many other distributed file systems exist: *Red Hat Cluster File System, Ceph File System,* and *Tachyon File System,* to name but three.

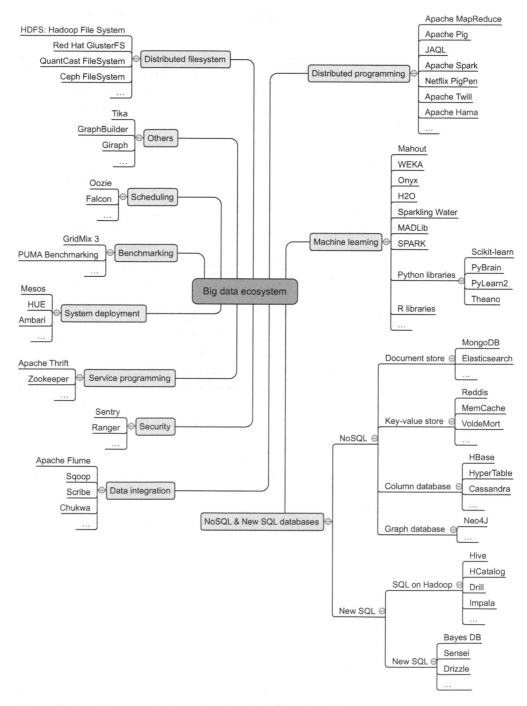

Figure 1.6 Big data technologies can be classified into a few main components.

1.4.2 *Distributed programming framework*

Once you have the data stored on the distributed file system, you want to exploit it. One important aspect of working on a distributed hard disk is that you won't move your data to your program, but rather you'll move your program to the data. When you start from scratch with a normal general-purpose programming language such as C, Python, or Java, you need to deal with the complexities that come with distributed programming, such as restarting jobs that have failed, tracking the results from the different subprocesses, and so on. Luckily, the open source community has developed many frameworks to handle this for you, and these give you a much better experience working with distributed data and dealing with many of the challenges it carries.

1.4.3 *Data integration framework*

Once you have a distributed file system in place, you need to add data. You need to move data from one source to another, and this is where the data integration frameworks such as Apache Sqoop and Apache Flume excel. The process is similar to an extract, transform, and load process in a traditional data warehouse.

1.4.4 *Machine learning frameworks*

When you have the data in place, it's time to extract the coveted insights. This is where you rely on the fields of machine learning, statistics, and applied mathematics. Before World War II everything needed to be calculated by hand, which severely limited the possibilities of data analysis. After World War II computers and scientific computing were developed. A single computer could do all the counting and calculations and a world of opportunities opened. Ever since this breakthrough, people only need to derive the mathematical formulas, write them in an algorithm, and load their data. With the enormous amount of data available nowadays, one computer can no longer handle the workload by itself. In fact, several algorithms developed in the previous millennium would never terminate before the end of the universe, even if you could use every computer available on Earth. This has to do with time complexity (https://en.wikipedia.org/wiki/Time_complexity). An example is trying to break a password by testing every possible combination. An example can be found at http://stackoverflow.com/questions/7055652/real-world-example-of-exponential-time-complexity. One of the biggest issues with the old algorithms is that they don't scale well. With the amount of data we need to analyze today, this becomes problematic, and specialized frameworks and libraries are required to deal with this amount of data. The most popular machine-learning library for Python is Scikit-learn. It's a great machine-learning toolbox, and we'll use it later in the book. There are, of course, other Python libraries:

- *PyBrain for neural networks*—Neural networks are learning algorithms that mimic the human brain in learning mechanics and complexity. Neural networks are often regarded as advanced and black box.

- *NLTK or Natural Language Toolkit*—As the name suggests, its focus is working with natural language. It's an extensive library that comes bundled with a number of text corpuses to help you model your own data.
- *Pylearn2*—Another machine learning toolbox but a bit less mature than Scikit-learn.
- *TensorFlow*—A Python library for deep learning provided by Google.

The landscape doesn't end with Python libraries, of course. Spark is a new Apache-licensed machine-learning engine, specializing in real-learn-time machine learning. It's worth taking a look at and you can read more about it at http://spark.apache.org/.

1.4.5 NoSQL databases

If you need to store huge amounts of data, you require software that's specialized in managing and querying this data. Traditionally this has been the playing field of relational databases such as Oracle SQL, MySQL, Sybase IQ, and others. While they're still the go-to technology for many use cases, new types of databases have emerged under the grouping of NoSQL databases.

The name of this group can be misleading, as "No" in this context stands for "Not Only." A lack of functionality in SQL isn't the biggest reason for the paradigm shift, and many of the NoSQL databases have implemented a version of SQL themselves. But traditional databases had shortcomings that didn't allow them to scale well. By solving several of the problems of traditional databases, NoSQL databases allow for a virtually endless growth of data. These shortcomings relate to every property of big data: their storage or processing power can't scale beyond a single node and they have no way to handle streaming, graph, or unstructured forms of data.

Many different types of databases have arisen, but they can be categorized into the following types:

- *Column databases*—Data is stored in columns, which allows algorithms to perform much faster queries. Newer technologies use cell-wise storage. Table-like structures are still important.
- *Document stores*—Document stores no longer use tables, but store every observation in a document. This allows for a much more flexible data scheme.
- *Streaming data*—Data is collected, transformed, and aggregated not in batches but in real time. Although we've categorized it here as a database to help you in tool selection, it's more a particular type of problem that drove creation of technologies such as Storm.
- *Key-value stores*—Data isn't stored in a table; rather you assign a key for every value, such as org.marketing.sales.2015: 20000. This scales well but places almost all the implementation on the developer.
- *SQL on Hadoop*—Batch queries on Hadoop are in a SQL-like language that uses the map-reduce framework in the background.
- *New SQL*—This class combines the scalability of NoSQL databases with the advantages of relational databases. They all have a SQL interface and a relational data model.

- *Graph databases*—Not every problem is best stored in a table. Particular problems are more naturally translated into graph theory and stored in graph databases. A classic example of this is a social network.

1.4.6   Scheduling tools

Scheduling tools help you automate repetitive tasks and trigger jobs based on events such as adding a new file to a folder. These are similar to tools such as CRON on Linux but are specifically developed for big data. You can use them, for instance, to start a MapReduce task whenever a new dataset is available in a directory.

1.4.7   Benchmarking tools

This class of tools was developed to optimize your big data installation by providing standardized profiling suites. A profiling suite is taken from a representative set of big data jobs. Benchmarking and optimizing the big data infrastructure and configuration aren't often jobs for data scientists themselves but for a professional specialized in setting up IT infrastructure; thus they aren't covered in this book. Using an optimized infrastructure can make a big cost difference. For example, if you can gain 10% on a cluster of 100 servers, you save the cost of 10 servers.

1.4.8   System deployment

Setting up a big data infrastructure isn't an easy task and assisting engineers in deploying new applications into the big data cluster is where system deployment tools shine. They largely automate the installation and configuration of big data components. This isn't a core task of a data scientist.

1.4.9   Service programming

Suppose that you've made a world-class soccer prediction application on Hadoop, and you want to allow others to use the predictions made by your application. However, you have no idea of the architecture or technology of everyone keen on using your predictions. Service tools excel here by exposing big data applications to other applications as a service. Data scientists sometimes need to expose their models through services. The best-known example is the REST service; REST stands for representational state transfer. It's often used to feed websites with data.

1.4.10   Security

Do you want everybody to have access to all of your data? You probably need to have fine-grained control over the access to data but don't want to manage this on an application-by-application basis. Big data security tools allow you to have central and fine-grained control over access to the data. Big data security has become a topic in its own right, and data scientists are usually only confronted with it as data consumers; seldom will they implement the security themselves. In this book we don't describe how to set up security on big data because this is a job for the security expert.

1.5 An introductory working example of Hadoop

We'll end this chapter with a small application in a big data context. For this we'll use a Hortonworks Sandbox image. This is a virtual machine created by Hortonworks to try some big data applications on a local machine. Later on in this book you'll see how Juju eases the installation of Hadoop on multiple machines.

We'll use a small data set of job salary data to run our first sample, but querying a large data set of billions of rows would be equally easy. The query language will seem like SQL, but behind the scenes a MapReduce job will run and produce a straightforward table of results, which can then be turned into a bar graph. The end result of this exercise looks like figure 1.7.

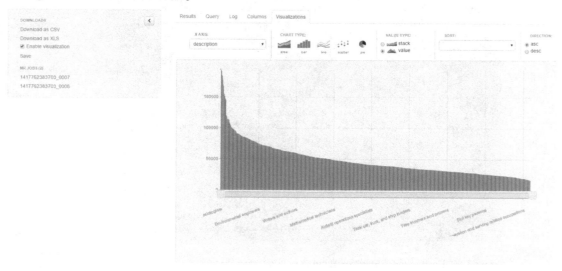

Figure 1.7 The end result: the average salary by job description

To get up and running as fast as possible we use a Hortonworks Sandbox inside Virtual-Box. VirtualBox is a virtualization tool that allows you to run another operating system inside your own operating system. In this case you can run CentOS with an existing Hadoop installation inside your installed operating system.

A few steps are required to get the sandbox up and running on VirtualBox. Caution, the following steps were applicable at the time this chapter was written (February 2015):

1 Download the virtual image from http://hortonworks.com/products/hortonworks-sandbox/#install.
2 Start your virtual machine host. VirtualBox can be downloaded from https://www.virtualbox.org/wiki/Downloads.

3 Press CTRL+I and select the virtual image from Hortonworks.
4 Click Next.
5 Click Import; after a little time your image should be imported.
6 Now select your virtual machine and click Run.
7 Give it a little time to start the CentOS distribution with the Hadoop installation running, as shown in figure 1.8. Notice the Sandbox version here is 2.1. With other versions things could be slightly different.

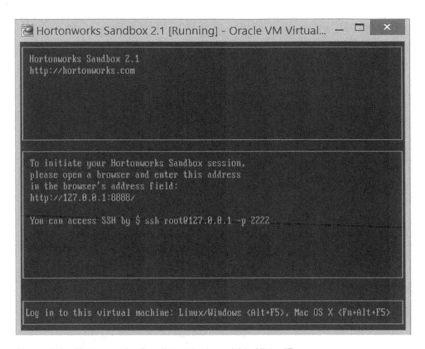

Figure 1.8 Hortonworks Sandbox running within VirtualBox

You can directly log on to the machine or use SSH to log on. For this application you'll use the web interface. Point your browser to the address http://127.0.0.1:8000 and you'll be welcomed with the screen shown in figure 1.9.

Hortonworks has uploaded two sample sets, which you can see in HCatalog. Just click the HCat button on the screen and you'll see the tables available to you (figure 1.10).

Figure 1.9   The Hortonworks Sandbox welcome screen available at http://127.0.0.1:8000

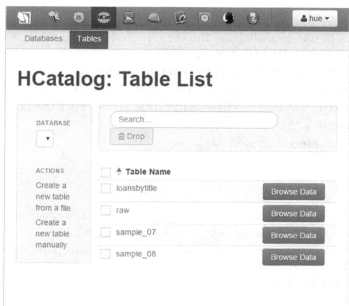

Figure 1.10   A list of available tables in HCatalog

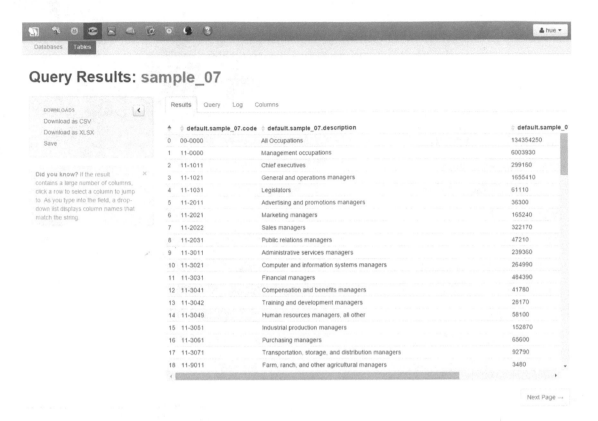

Figure 1.11   The contents of the table

To see the contents of the data, click the Browse Data button next to the sample_07 entry to get the next screen (figure 1.11).

This looks like an ordinary table, and Hive is a tool that lets you approach it like an ordinary database with SQL. That's right: in Hive you get your results using HiveQL, a dialect of plain-old SQL. To open the Beeswax HiveQL editor, click the Beeswax button in the menu (figure 1.12).

To get your results, execute the following query:

```
Select description, avg(salary) as average_salary from sample_07 group by
description order by average_salary desc.
```

Click the Execute button. Hive translates your HiveQL into a MapReduce job and executes it in your Hadoop environment, as you can see in figure 1.13.

Best however to avoid reading the log window for now. At this point, it's misleading. If this is your first query, then it could take 30 seconds. Hadoop is famous for its warming periods. That discussion is for later, though.

Figure 1.12   **You can execute a HiveQL command in the Beeswax HiveQL editor. Behind the scenes it's translated into a MapReduce job.**

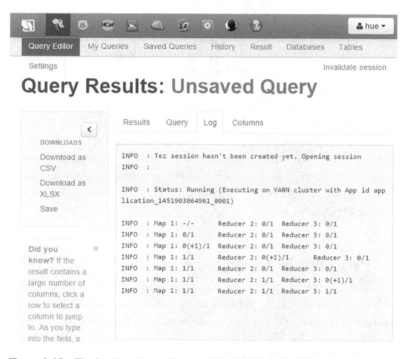

Figure 1.13   **The logging shows that your HiveQL is translated into a MapReduce job. Note: This log was from the February 2015 version of HDP, so the current version might look slightly different.**

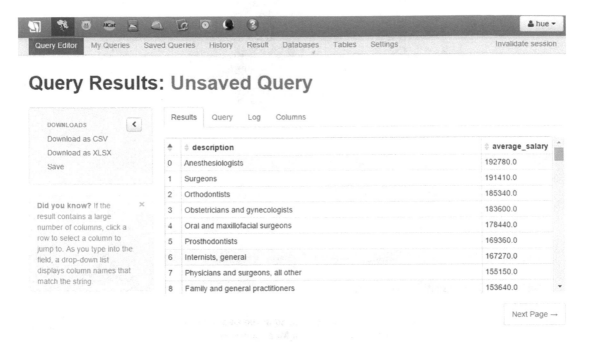

Figure 1.14   The end result: an overview of the average salary by profession

After a while the result appears. Great work! The conclusion of this, as shown in figure 1.14, is that going to medical school is a good investment. Surprised?

With this table we conclude our introductory Hadoop tutorial.

Although this chapter was but the beginning, it might have felt a bit overwhelming at times. It's recommended to leave it be for now and come back here again when all the concepts have been thoroughly explained. Data science is a broad field so it comes with a broad vocabulary. We hope to give you a glimpse of most of it during our time together. Afterward, you pick and choose and hone your skills in whatever direction interests you the most. That's what "Introducing Data Science" is all about and we hope you'll enjoy the ride with us.

1.6   *Summary*

In this chapter you learned the following:

- *Big data* is a blanket term for any collection of data sets so large or complex that it becomes difficult to process them using traditional data management techniques. They are characterized by the four Vs: velocity, variety, volume, and veracity.
- *Data science* involves using methods to analyze small data sets to the gargantuan ones big data is all about.

- Even though the *data science process* isn't linear it can be divided into steps:
  1 Setting the research goal
  2 Gathering data
  3 Data preparation
  4 Data exploration
  5 Modeling
  6 Presentation and automation
- The big data landscape is more than Hadoop alone. It consists of many different technologies that can be categorized into the following:
  - File system
  - Distributed programming frameworks
  - Data integration
  - Databases
  - Machine learning
  - Security
  - Scheduling
  - Benchmarking
  - System deployment
  - Service programming
- Not every big data category is utilized heavily by data scientists. They focus mainly on the file system, the distributed programming frameworks, databases, and machine learning. They do come in contact with the other components, but these are domains of other professions.
- Data can come in different forms. The main forms are
  - Structured data
  - Unstructured data
  - Natural language data
  - Machine data
  - Graph-based data
  - Streaming data

The data science process

This chapter covers

- Understanding the flow of a data science process
- Discussing the steps in a data science process

The goal of this chapter is to give an overview of the data science process without diving into big data yet. You'll learn how to work with big data sets, streaming data, and text data in subsequent chapters.

2.1 Overview of the data science process

Following a structured approach to data science helps you to maximize your chances of success in a data science project at the lowest cost. It also makes it possible to take up a project as a team, with each team member focusing on what they do best. Take care, however: this approach may not be suitable for every type of project or be the only way to do good data science.

The typical data science process consists of six steps through which you'll iterate, as shown in figure 2.1.

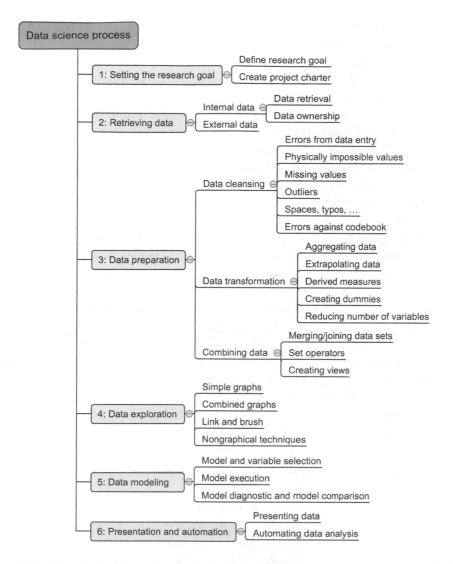

Figure 2.1   The six steps of the data science process

Figure 2.1 summarizes the data science process and shows the main steps and actions you'll take during a project. The following list is a short introduction; each of the steps will be discussed in greater depth throughout this chapter.

1   The first step of this process is setting a *research goal*. The main purpose here is making sure all the stakeholders understand the *what, how,* and *why* of the project. In every serious project this will result in a project charter.

2   The second phase is *data retrieval.* You want to have data available for analysis, so this step includes finding suitable data and getting access to the data from the

data owner. The result is data in its raw form, which probably needs polishing and transformation before it becomes usable.

3   Now that you have the raw data, it's time to *prepare* it. This includes transforming the data from a raw form into data that's directly usable in your models. To achieve this, you'll detect and correct different kinds of errors in the data, combine data from different data sources, and transform it. If you have successfully completed this step, you can progress to data visualization and modeling.

4   The fourth step is *data exploration*. The goal of this step is to gain a deep understanding of the data. You'll look for patterns, correlations, and deviations based on visual and descriptive techniques. The insights you gain from this phase will enable you to start modeling.

5   Finally, we get to the sexiest part: *model building* (often referred to as "data modeling" throughout this book). It is now that you attempt to gain the insights or make the predictions stated in your project charter. Now is the time to bring out the heavy guns, but remember research has taught us that often (but not always) a combination of simple models tends to outperform one complicated model. If you've done this phase right, you're almost done.

6   The last step of the data science model is *presenting your results and automating the analysis,* if needed. One goal of a project is to change a process and/or make better decisions. You may still need to convince the business that your findings will indeed change the business process as expected. This is where you can shine in your influencer role. The importance of this step is more apparent in projects on a strategic and tactical level. Certain projects require you to perform the business process over and over again, so automating the project will save time.

In reality you won't progress in a linear way from step 1 to step 6. Often you'll regress and iterate between the different phases.

Following these six steps pays off in terms of a higher project success ratio and increased impact of research results. This process ensures you have a well-defined research plan, a good understanding of the business question, and clear deliverables before you even start looking at data. The first steps of your process focus on getting high-quality data as input for your models. This way your models will perform better later on. In data science there's a well-known saying: *Garbage in equals garbage out.*

Another benefit of following a structured approach is that you work more in *prototype mode* while you search for the best model. When building a *prototype,* you'll probably try multiple models and won't focus heavily on issues such as program speed or writing code against standards. This allows you to focus on bringing business value instead.

Not every project is initiated by the business itself. Insights learned during analysis or the arrival of new data can spawn new projects. When the data science team generates an idea, work has already been done to make a proposition and find a business sponsor.

Dividing a project into smaller stages also allows employees to work together as a team. It's impossible to be a specialist in everything. You'd need to know how to upload all the data to all the different databases, find an optimal data scheme that works not only for your application but also for other projects inside your company, and then keep track of all the statistical and data-mining techniques, while also being an expert in presentation tools and business politics. That's a hard task, and it's why more and more companies rely on a team of specialists rather than trying to find one person who can do it all.

The process we described in this section is best suited for a data science project that contains only a few models. It's not suited for every type of project. For instance, a project that contains millions of real-time models would need a different approach than the flow we describe here. A beginning data scientist should get a long way following this manner of working, though.

2.1.1 Don't be a slave to the process

Not every project will follow this blueprint, because your process is subject to the preferences of the data scientist, the company, and the nature of the project you work on. Some companies may require you to follow a strict protocol, whereas others have a more informal manner of working. In general, you'll need a structured approach when you work on a complex project or when many people or resources are involved.

The *agile* project model is an alternative to a sequential process with iterations. As this methodology wins more ground in the IT department and throughout the company, it's also being adopted by the data science community. Although the agile methodology is suitable for a data science project, many company policies will favor a more rigid approach toward data science.

Planning every detail of the data science process upfront isn't always possible, and more often than not you'll iterate between the different steps of the process. For instance, after the briefing you start your normal flow until you're in the exploratory data analysis phase. Your graphs show a distinction in the behavior between two groups—men and women maybe? You aren't sure because you don't have a variable that indicates whether the customer is male or female. You need to retrieve an extra data set to confirm this. For this you need to go through the approval process, which indicates that you (or the business) need to provide a kind of project charter. In big companies, getting all the data you need to finish your project can be an ordeal.

2.2 Step 1: Defining research goals and creating a project charter

A project starts by understanding the *what*, the *why*, and the *how* of your project (figure 2.2). What does the company expect you to do? And why does management place such a value on your research? Is it part of a bigger strategic picture or a "lone wolf" project originating from an opportunity someone detected? Answering these three

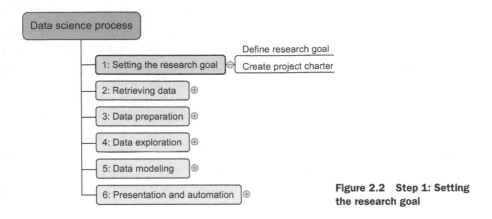

Figure 2.2   **Step 1: Setting the research goal**

questions (what, why, how) is the goal of the first phase, so that everybody knows what to do and can agree on the best course of action.

The outcome should be a clear research goal, a good understanding of the context, well-defined deliverables, and a plan of action with a timetable. This information is then best placed in a project charter. The length and formality can, of course, differ between projects and companies. In this early phase of the project, people skills and business acumen are more important than great technical prowess, which is why this part will often be guided by more senior personnel.

2.2.1   *Spend time understanding the goals and context of your research*

An essential outcome is the research goal that states the purpose of your assignment in a clear and focused manner. Understanding the business goals and context is critical for project success. Continue asking questions and devising examples until you grasp the exact business expectations, identify how your project fits in the bigger picture, appreciate how your research is going to change the business, and understand how they'll use your results. Nothing is more frustrating than spending months researching something until you have that one moment of brilliance and solve the problem, but when you report your findings back to the organization, everyone immediately realizes that you misunderstood their question. Don't skim over this phase lightly. Many data scientists fail here: despite their mathematical wit and scientific brilliance, they never seem to grasp the business goals and context.

2.2.2   *Create a project charter*

Clients like to know upfront what they're paying for, so after you have a good understanding of the business problem, try to get a formal agreement on the deliverables. All this information is best collected in a project charter. For any significant project this would be mandatory.

A project charter requires teamwork, and your input covers at least the following:

- A clear research goal
- The project mission and context
- How you're going to perform your analysis
- What resources you expect to use
- Proof that it's an achievable project, or proof of concepts
- Deliverables and a measure of success
- A timeline

Your client can use this information to make an estimation of the project costs and the data and people required for your project to become a success.

2.3 Step 2: Retrieving data

The next step in data science is to retrieve the required data (figure 2.3). Sometimes you need to go into the field and design a data collection process yourself, but most of the time you won't be involved in this step. Many companies will have already collected and stored the data for you, and what they don't have can often be bought from third parties. Don't be afraid to look outside your organization for data, because more and more organizations are making even high-quality data freely available for public and commercial use.

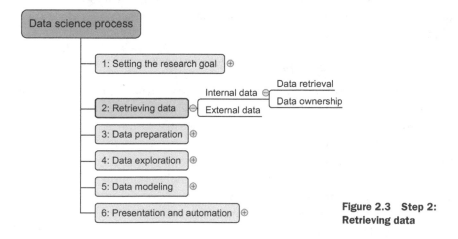

Figure 2.3 Step 2: Retrieving data

Data can be stored in many forms, ranging from simple text files to tables in a database. The objective now is acquiring all the data you need. This may be difficult, and even if you succeed, data is often like a diamond in the rough: it needs polishing to be of any use to you.

2.3.1   *Start with data stored within the company*

Your first act should be to assess the relevance and quality of the data that's readily available within your company. Most companies have a program for maintaining key data, so much of the cleaning work may already be done. This data can be stored in official data repositories such as *databases, data marts, data warehouses,* and *data lakes* maintained by a team of IT professionals. The primary goal of a database is data storage, while a data warehouse is designed for reading and analyzing that data. A data mart is a subset of the data warehouse and geared toward serving a specific business unit. While data warehouses and data marts are home to preprocessed data, data lakes contains data in its natural or raw format. But the possibility exists that your data still resides in Excel files on the desktop of a domain expert.

Finding data even within your own company can sometimes be a challenge. As companies grow, their data becomes scattered around many places. Knowledge of the data may be dispersed as people change positions and leave the company. Documentation and metadata aren't always the top priority of a delivery manager, so it's possible you'll need to develop some Sherlock Holmes–like skills to find all the lost bits.

Getting access to data is another difficult task. Organizations understand the value and sensitivity of data and often have policies in place so everyone has access to what they need and nothing more. These policies translate into physical and digital barriers called *Chinese walls.* These "walls" are mandatory and well-regulated for customer data in most countries. This is for good reasons, too; imagine everybody in a credit card company having access to your spending habits. Getting access to the data may take time and involve company politics.

2.3.2   *Don't be afraid to shop around*

If data isn't available inside your organization, look outside your organization's walls. Many companies specialize in collecting valuable information. For instance, Nielsen and GFK are well known for this in the retail industry. Other companies provide data so that you, in turn, can enrich their services and ecosystem. Such is the case with Twitter, LinkedIn, and Facebook.

Although data is considered an asset more valuable than oil by certain companies, more and more governments and organizations share their data for free with the world. This data can be of excellent quality; it depends on the institution that creates and manages it. The information they share covers a broad range of topics such as the number of accidents or amount of drug abuse in a certain region and its demographics. This data is helpful when you want to enrich proprietary data but also convenient when training your data science skills at home. Table 2.1 shows only a small selection from the growing number of open-data providers.

Table 2.1 A list of open-data providers that should get you started

Open data site	Description
Data.gov	The home of the US Government's open data
https://open-data.europa.eu/	The home of the European Commission's open data
Freebase.org	An open database that retrieves its information from sites like Wikipedia, MusicBrains, and the SEC archive
Data.worldbank.org	Open data initiative from the World Bank
Aiddata.org	Open data for international development
Open.fda.gov	Open data from the US Food and Drug Administration

2.3.3 *Do data quality checks now to prevent problems later*

Expect to spend a good portion of your project time doing data correction and cleansing, sometimes up to 80%. The retrieval of data is the first time you'll inspect the data in the data science process. Most of the errors you'll encounter during the data-gathering phase are easy to spot, but being too careless will make you spend many hours solving data issues that could have been prevented during data import.

You'll investigate the data during the import, data preparation, and exploratory phases. The difference is in the goal and the depth of the investigation. During *data retrieval*, you check to see if the data is equal to the data in the source document and look to see if you have the right data types. This shouldn't take too long; when you have enough evidence that the data is similar to the data you find in the source document, you stop. With *data preparation*, you do a more elaborate check. If you did a good job during the previous phase, the errors you find now are also present in the source document. The focus is on the content of the variables: you want to get rid of typos and other data entry errors and bring the data to a common standard among the data sets. For example, you might correct USQ to USA and United Kingdom to UK. During the *exploratory phase* your focus shifts to what you can learn from the data. Now you assume the data to be clean and look at the statistical properties such as distributions, correlations, and outliers. You'll often iterate over these phases. For instance, when you discover outliers in the exploratory phase, they can point to a data entry error. Now that you understand how the quality of the data is improved during the process, we'll look deeper into the data preparation step.

2.4 *Step 3: Cleansing, integrating, and transforming data*

The data received from the data retrieval phase is likely to be "a diamond in the rough." Your task now is to sanitize and prepare it for use in the modeling and reporting phase. Doing so is tremendously important because your models will perform better and you'll lose less time trying to fix strange output. It can't be mentioned nearly enough times: garbage in equals garbage out. Your model needs the data in a specific

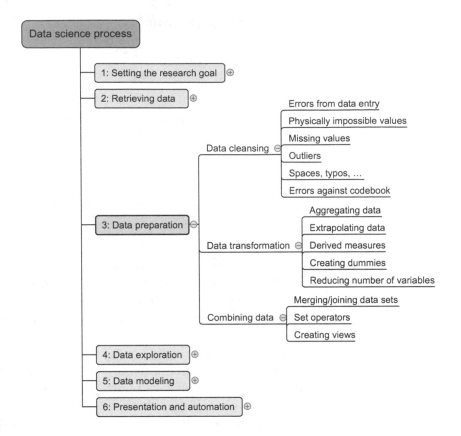

Figure 2.4   Step 3: Data preparation

format, so data transformation will always come into play. It's a good habit to correct data errors as early on in the process as possible. However, this isn't always possible in a realistic setting, so you'll need to take corrective actions in your program.

Figure 2.4 shows the most common actions to take during the data cleansing, integration, and transformation phase.

This mind map may look a bit abstract for now, but we'll handle all of these points in more detail in the next sections. You'll see a great commonality among all of these actions.

2.4.1   *Cleansing data*

Data cleansing is a subprocess of the data science process that focuses on removing errors in your data so your data becomes a true and consistent representation of the processes it originates from.

By "true and consistent representation" we imply that at least two types of errors exist. The first type is the *interpretation error*, such as when you take the value in your data for granted, like saying that a person's age is greater than 300 years. The second type of error points to *inconsistencies* between data sources or against your company's standardized values. An example of this class of errors is putting "Female" in one table and "F" in another when they represent the same thing: that the person is female. Another example is that you use Pounds in one table and Dollars in another. Too many possible errors exist for this list to be exhaustive, but table 2.2 shows an overview of the types of errors that can be detected with easy checks—the "low hanging fruit," as it were.

Table 2.2   An overview of common errors

General solution	
Try to fix the problem early in the data acquisition chain or else fix it in the program.	
Error description	**Possible solution**
Errors pointing to false values within one data set	
Mistakes during data entry	Manual overrules
Redundant white space	Use string functions
Impossible values	Manual overrules
Missing values	Remove observation or value
Outliers	Validate and, if erroneous, treat as missing value (remove or insert)
Errors pointing to inconsistencies between data sets	
Deviations from a code book	Match on keys or else use manual overrules
Different units of measurement	Recalculate
Different levels of aggregation	Bring to same level of measurement by aggregation or extrapolation

Sometimes you'll use more advanced methods, such as simple modeling, to find and identify data errors; diagnostic plots can be especially insightful. For example, in figure 2.5 we use a measure to identify data points that seem out of place. We do a regression to get acquainted with the data and detect the influence of individual observations on the regression line. When a single observation has too much influence, this can point to an error in the data, but it can also be a valid point. At the data cleansing stage, these advanced methods are, however, rarely applied and often regarded by certain data scientists as overkill.

Now that we've given the overview, it's time to explain these errors in more detail.

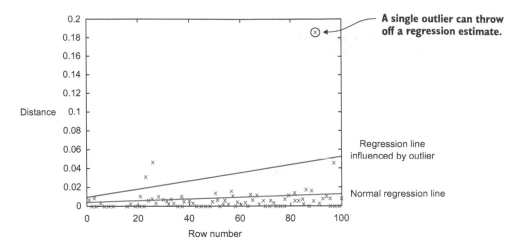

Figure 2.5   The encircled point influences the model heavily and is worth investigating because it can point to a region where you don't have enough data or might indicate an error in the data, but it also can be a valid data point.

DATA ENTRY ERRORS

Data collection and data entry are error-prone processes. They often require human intervention, and because humans are only human, they make typos or lose their concentration for a second and introduce an error into the chain. But data collected by machines or computers isn't free from errors either. Errors can arise from human sloppiness, whereas others are due to machine or hardware failure. Examples of errors originating from machines are transmission errors or bugs in the extract, transform, and load phase (ETL).

For small data sets you can check every value by hand. Detecting data errors when the variables you study don't have many classes can be done by tabulating the data with counts. When you have a variable that can take only two values: "Good" and "Bad", you can create a frequency table and see if those are truly the only two values present. In table 2.3, the values "Godo" and "Bade" point out something went wrong in at least 16 cases.

Table 2.3   Detecting outliers on simple variables with a frequency table

Value	Count
Good	1598647
Bad	1354468
Godo	15
Bade	1

Most errors of this type are easy to fix with simple assignment statements and if-then-else rules:

```
if x == "Godo":
    x = "Good"
if x == "Bade":
    x = "Bad"
```

REDUNDANT WHITESPACE

Whitespaces tend to be hard to detect but cause errors like other redundant characters would. Who hasn't lost a few days in a project because of a bug that was caused by whitespaces at the end of a string? You ask the program to join two keys and notice that observations are missing from the output file. After looking for days through the code, you finally find the bug. Then comes the hardest part: explaining the delay to the project stakeholders. The cleaning during the ETL phase wasn't well executed, and keys in one table contained a whitespace at the end of a string. This caused a mismatch of keys such as "FR " – "FR", dropping the observations that couldn't be matched.

If you know to watch out for them, fixing redundant whitespaces is luckily easy enough in most programming languages. They all provide string functions that will remove the leading and trailing whitespaces. For instance, in Python you can use the `strip()` function to remove leading and trailing spaces.

FIXING CAPITAL LETTER MISMATCHES　　Capital letter mismatches are common. Most programming languages make a distinction between "Brazil" and "brazil". In this case you can solve the problem by applying a function that returns both strings in lowercase, such as `.lower()` in Python. `"Brazil".lower() == "brazil".lower()` should result in `true`.

IMPOSSIBLE VALUES AND SANITY CHECKS

Sanity checks are another valuable type of data check. Here you check the value against physically or theoretically impossible values such as people taller than 3 meters or someone with an age of 299 years. Sanity checks can be directly expressed with rules:

```
check = 0 <= age <= 120
```

OUTLIERS

An outlier is an observation that seems to be distant from other observations or, more specifically, one observation that follows a different logic or generative process than the other observations. The easiest way to find outliers is to use a plot or a table with the minimum and maximum values. An example is shown in figure 2.6.

The plot on the top shows no outliers, whereas the plot on the bottom shows possible outliers on the upper side when a normal distribution is expected. The normal distribution, or Gaussian distribution, is the most common distribution in natural sciences.

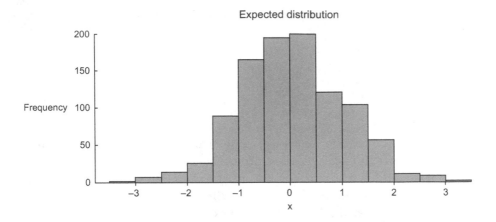

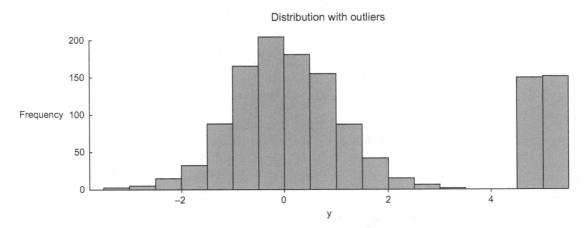

Figure 2.6   Distribution plots are helpful in detecting outliers and helping you understand the variable.

It shows most cases occurring around the average of the distribution and the occurrences decrease when further away from it. The high values in the bottom graph can point to outliers when assuming a normal distribution. As we saw earlier with the regression example, outliers can gravely influence your data modeling, so investigate them first.

DEALING WITH MISSING VALUES

Missing values aren't necessarily wrong, but you still need to handle them separately; certain modeling techniques can't handle missing values. They might be an indicator that something went wrong in your data collection or that an error happened in the ETL process. Common techniques data scientists use are listed in table 2.4.

Table 2.4   An overview of techniques to handle missing data

Technique	Advantage	Disadvantage
Omit the values	Easy to perform	You lose the information from an observation
Set value to `null`	Easy to perform	Not every modeling technique and/or implementation can handle `null` values
Impute a static value such as 0 or the mean	Easy to perform You don't lose information from the other variables in the observation	Can lead to false estimations from a model
Impute a value from an estimated or theoretical distribution	Does not disturb the model as much	Harder to execute You make data assumptions
Modeling the value (nondependent)	Does not disturb the model too much	Can lead to too much confidence in the model Can artificially raise dependence among the variables Harder to execute You make data assumptions

Which technique to use at what time is dependent on your particular case. If, for instance, you don't have observations to spare, omitting an observation is probably not an option. If the variable can be described by a stable distribution, you could impute based on this. However, maybe a missing value actually means "zero"? This can be the case in sales for instance: if no promotion is applied on a customer basket, that customer's promo is missing, but most likely it's also 0, no price cut.

DEVIATIONS FROM A CODE BOOK

Detecting errors in larger data sets against a code book or against standardized values can be done with the help of set operations. A code book is a description of your data, a form of metadata. It contains things such as the number of variables per observation, the number of observations, and what each encoding within a variable means. (For instance "0" equals "negative", "5" stands for "very positive".) A code book also tells the type of data you're looking at: is it hierarchical, graph, something else?

You look at those values that are present in set A but not in set B. These are values that should be corrected. It's no coincidence that *sets* are the data structure that we'll use when we're working in code. It's a good habit to give your data structures additional thought; it can save work and improve the performance of your program.

If you have multiple values to check, it's better to put them from the code book into a table and use a difference operator to check the discrepancy between both tables. This way, you can profit from the power of a database directly. More on this in chapter 5.

DIFFERENT UNITS OF MEASUREMENT

When integrating two data sets, you have to pay attention to their respective units of measurement. An example of this would be when you study the prices of gasoline in the world. To do this you gather data from different data providers. Data sets can contain prices per gallon and others can contain prices per liter. A simple conversion will do the trick in this case.

DIFFERENT LEVELS OF AGGREGATION

Having different levels of aggregation is similar to having different types of measurement. An example of this would be a data set containing data per week versus one containing data per work week. This type of error is generally easy to detect, and *summarizing* (or the inverse, *expanding*) the data sets will fix it.

After cleaning the data errors, you combine information from different data sources. But before we tackle this topic we'll take a little detour and stress the importance of cleaning data as early as possible.

2.4.2   *Correct errors as early as possible*

A good practice is to mediate data errors as early as possible in the data collection chain and to fix as little as possible inside your program while fixing the origin of the problem. Retrieving data is a difficult task, and organizations spend millions of dollars on it in the hope of making better decisions. The data collection process is error-prone, and in a big organization it involves many steps and teams.

Data should be cleansed when acquired for many reasons:

- Not everyone spots the data anomalies. Decision-makers may make costly mistakes on information based on incorrect data from applications that fail to correct for the faulty data.
- If errors are not corrected early on in the process, the cleansing will have to be done for every project that uses that data.
- Data errors may point to a business process that isn't working as designed. For instance, both authors worked at a retailer in the past, and they designed a couponing system to attract more people and make a higher profit. During a data science project, we discovered clients who abused the couponing system and earned money while purchasing groceries. The goal of the couponing system was to stimulate cross-selling, not to give products away for free. This flaw cost the company money and nobody in the company was aware of it. In this case the data wasn't technically wrong but came with unexpected results.
- Data errors may point to defective equipment, such as broken transmission lines and defective sensors.
- Data errors can point to bugs in software or in the integration of software that may be critical to the company. While doing a small project at a bank we discovered that two software applications used different local settings. This caused problems with numbers greater than 1,000. For one app the number 1.000 meant one, and for the other it meant one thousand.

Fixing the data as soon as it's captured is nice in a perfect world. Sadly, a data scientist doesn't always have a say in the data collection and simply telling the IT department to fix certain things may not make it so. If you can't correct the data at the source, you'll need to handle it inside your code. Data manipulation doesn't end with correcting mistakes; you still need to combine your incoming data.

As a final remark: always keep a copy of your original data (if possible). Sometimes you start cleaning data but you'll make mistakes: impute variables in the wrong way, delete outliers that had interesting additional information, or alter data as the result of an initial misinterpretation. If you keep a copy you get to try again. For "flowing data" that's manipulated at the time of arrival, this isn't always possible and you'll have accepted a period of tweaking before you get to use the data you are capturing. One of the more difficult things isn't the data cleansing of individual data sets however, it's combining different sources into a whole that makes more sense.

2.4.3 Combining data from different data sources

Your data comes from several different places, and in this substep we focus on integrating these different sources. Data varies in size, type, and structure, ranging from databases and Excel files to text documents.

We focus on data in table structures in this chapter for the sake of brevity. It's easy to fill entire books on this topic alone, and we choose to focus on the data science process instead of presenting scenarios for every type of data. But keep in mind that other types of data sources exist, such as key-value stores, document stores, and so on, which we'll handle in more appropriate places in the book.

THE DIFFERENT WAYS OF COMBINING DATA

You can perform two operations to combine information from different data sets. The first operation is *joining*: enriching an observation from one table with information from another table. The second operation is *appending* or *stacking*: adding the observations of one table to those of another table.

When you combine data, you have the option to create a new physical table or a virtual table by creating a view. The advantage of a view is that it doesn't consume more disk space. Let's elaborate a bit on these methods.

JOINING TABLES

Joining tables allows you to combine the information of one observation found in one table with the information that you find in another table. The focus is on enriching a single observation. Let's say that the first table contains information about the purchases of a customer and the other table contains information about the region where your customer lives. Joining the tables allows you to combine the information so that you can use it for your model, as shown in figure 2.7.

To join tables, you use variables that represent the same object in both tables, such as a date, a country name, or a Social Security number. These common fields are known as keys. When these keys also uniquely define the records in the table they

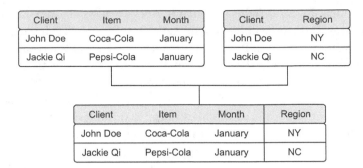

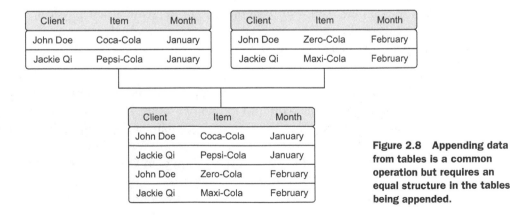

Figure 2.7   Joining two tables on the Item and Region keys

are called *primary keys*. One table may have buying behavior and the other table may have demographic information on a person. In figure 2.7 both tables contain the client name, and this makes it easy to enrich the client expenditures with the region of the client. People who are acquainted with Excel will notice the similarity with using a lookup function.

The number of resulting rows in the output table depends on the exact join type that you use. We introduce the different types of joins later in the book.

APPENDING TABLES

Appending or stacking tables is effectively adding observations from one table to another table. Figure 2.8 shows an example of appending tables. One table contains the observations from the month January and the second table contains observations from the month February. The result of appending these tables is a larger one with the observations from January as well as February. The equivalent operation in set theory would be the union, and this is also the command in SQL, the common language of relational databases. Other set operators are also used in data science, such as set difference and intersection.

Figure 2.8   Appending data from tables is a common operation but requires an equal structure in the tables being appended.

USING VIEWS TO SIMULATE DATA JOINS AND APPENDS

To avoid duplication of data, you virtually combine data with views. In the previous example we took the monthly data and combined it in a new physical table. The problem is that we duplicated the data and therefore needed more storage space. In the example we're working with, that may not cause problems, but imagine that every table consists of terabytes of data; then it becomes problematic to duplicate the data. For this reason, the concept of a view was invented. A view behaves as if you're working on a table, but this table is nothing but a virtual layer that combines the tables for you. Figure 2.9 shows how the sales data from the different months is combined virtually into a yearly sales table instead of duplicating the data. Views do come with a drawback, however. While a table join is only performed once, the join that creates the view is recreated every time it's queried, using more processing power than a pre-calculated table would have.

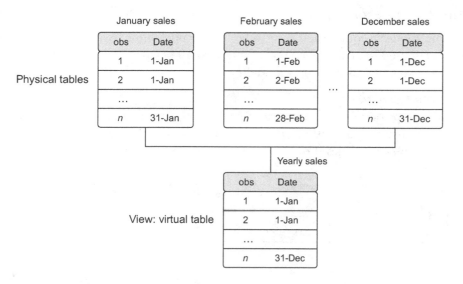

Figure 2.9   A view helps you combine data without replication.

ENRICHING AGGREGATED MEASURES

Data enrichment can also be done by adding calculated information to the table, such as the total number of sales or what percentage of total stock has been sold in a certain region (figure 2.10).

Extra measures such as these can add perspective. Looking at figure 2.10, we now have an aggregated data set, which in turn can be used to calculate the participation of each product within its category. This could be useful during data exploration but more so when creating data models. As always this depends on the exact case, but from our experience models with "relative measures" such as % sales (quantity of

Product class	Product	Sales in $	Sales t-1 in $	Growth	Sales by product class	Rank sales
A	B	X	Y	(X-Y) / Y	AX	NX
Sport	Sport 1	95	98	−3.06%	215	2
Sport	Sport 2	120	132	−9.09%	215	1
Shoes	Shoes 1	10	6	66.67%	10	3

Figure 2.10   Growth, sales by product class, and rank sales are examples of derived and aggregate measures.

product sold/total quantity sold) tend to outperform models that use the raw numbers (quantity sold) as input.

2.4.4   *Transforming data*

Certain models require their data to be in a certain shape. Now that you've cleansed and integrated the data, this is the next task you'll perform: transforming your data so it takes a suitable form for data modeling.

TRANSFORMING DATA

Relationships between an input variable and an output variable aren't always linear. Take, for instance, a relationship of the form $y = ae^{bx}$. Taking the log of the independent variables simplifies the estimation problem dramatically. Figure 2.11 shows how

x	1	2	3	4	5	6	7	8	9	10
log(x)	0.00	0.43	0.68	0.86	1.00	1.11	1.21	1.29	1.37	1.43
y	0.00	0.44	0.69	0.87	1.02	1.11	1.24	1.32	1.38	1.46

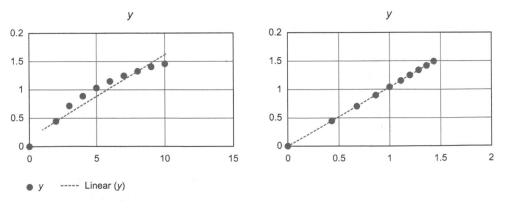

Figure 2.11   Transforming x to log x makes the relationship between x and y linear (right), compared with the non-log x (left).

transforming the input variables greatly simplifies the estimation problem. Other times you might want to combine two variables into a new variable.

REDUCING THE NUMBER OF VARIABLES

Sometimes you have too many variables and need to reduce the number because they don't add new information to the model. Having too many variables in your model makes the model difficult to handle, and certain techniques don't perform well when you overload them with too many input variables. For instance, all the techniques based on a Euclidean distance perform well only up to 10 variables.

Euclidean distance

Euclidean distance or "ordinary" distance is an extension to one of the first things anyone learns in mathematics about triangles (trigonometry): Pythagoras's leg theorem. If you know the length of the two sides next to the 90° angle of a right-angled triangle you can easily derive the length of the remaining side (hypotenuse). The formula for this is hypotenuse $= \sqrt{(side1 + side2)^2}$. The Euclidean distance between two points in a two-dimensional plane is calculated using a similar formula: distance $= \sqrt{((x1 - x2)^2 + (y1 - y2)^2)}$. If you want to expand this distance calculation to more dimensions, add the coordinates of the point within those higher dimensions to the formula. For three dimensions we get distance $= \sqrt{((x1 - x2)^2 + (y1 - y2)^2 + (z1 - z2)^2)}$.

Data scientists use special methods to reduce the number of variables but retain the maximum amount of data. We'll discuss several of these methods in chapter 3. Figure 2.12 shows how reducing the number of variables makes it easier to understand the

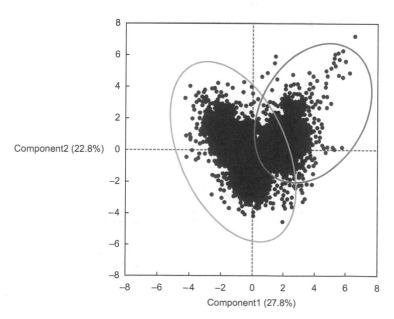

Figure 2.12 Variable reduction allows you to reduce the number of variables while maintaining as much information as possible.

key values. It also shows how two variables account for 50.6% of the variation within the data set (component1 = 27.8% + component2 = 22.8%). These variables, called "component1" and "component2," are both combinations of the original variables. They're the *principal components* of the underlying data structure. If it isn't all that clear at this point, don't worry, principal components analysis (PCA) will be explained more thoroughly in chapter 3. What you can also see is the presence of a third (unknown) variable that splits the group of observations into two.

TURNING VARIABLES INTO DUMMIES

Variables can be turned into dummy variables (figure 2.13). *Dummy variables* can only take two values: true(1) or false(0). They're used to indicate the absence of a categorical effect that may explain the observation. In this case you'll make separate columns for the classes stored in one variable and indicate it with 1 if the class is present and 0 otherwise. An example is turning one column named Weekdays into the columns Monday through Sunday. You use an indicator to show if the observation was on a Monday; you put 1 on Monday and 0 elsewhere. Turning variables into dummies is a technique that's used in modeling and is popular with, but not exclusive to, economists.

In this section we introduced the third step in the data science process—cleaning, transforming, and integrating data—which changes your raw data into usable input for the modeling phase. The next step in the data science process is to get a better

Customer	Year	Gender	Sales
1	2015	F	10
2	2015	M	8
1	2016	F	11
3	2016	M	12
4	2017	F	14
3	2017	M	13

M   F

Customer	Year	Sales	Male	Female
1	2015	10	0	1
1	2016	11	0	1
2	2015	8	1	0
3	2016	12	1	0
3	2017	13	1	0
4	2017	14	0	1

Figure 2.13   Turning variables into dummies is a data transformation that breaks a variable that has multiple classes into multiple variables, each having only two possible values: 0 or 1.

understanding of the content of the data and the relationships between the variables and observations; we explore this in the next section.

2.5   *Step 4: Exploratory data analysis*

During exploratory data analysis you take a deep dive into the data (see figure 2.14). Information becomes much easier to grasp when shown in a picture, therefore you mainly use graphical techniques to gain an understanding of your data and the interactions between variables. This phase is about exploring data, so keeping your mind open and your eyes peeled is essential during the exploratory data analysis phase. The goal isn't to cleanse the data, but it's common that you'll still discover anomalies you missed before, forcing you to take a step back and fix them.

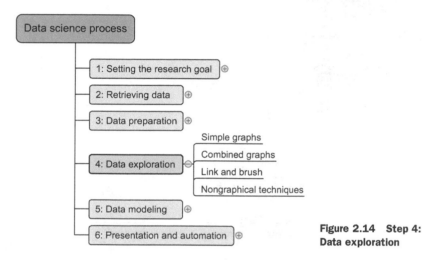

Figure 2.14   Step 4: Data exploration

The visualization techniques you use in this phase range from simple line graphs or histograms, as shown in figure 2.15, to more complex diagrams such as Sankey and network graphs. Sometimes it's useful to compose a composite graph from simple graphs to get even more insight into the data. Other times the graphs can be animated or made interactive to make it easier and, let's admit it, way more fun. An example of an interactive Sankey diagram can be found at http://bost.ocks.org/mike/sankey/.

Mike Bostock has interactive examples of almost any type of graph. It's worth spending time on his website, though most of his examples are more useful for data presentation than data exploration.

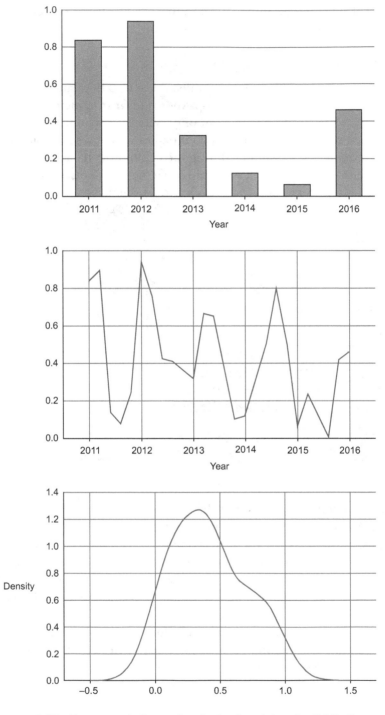

Figure 2.15   From top to bottom, a bar chart, a line plot, and a distribution are some of the graphs used in exploratory analysis.

These plots can be combined to provide even more insight, as shown in figure 2.16.

Overlaying several plots is common practice. In figure 2.17 we combine simple graphs into a Pareto diagram, or 80-20 diagram.

Figure 2.18 shows another technique: *brushing and linking*. With brushing and linking you combine and link different graphs and tables (or views) so changes in one graph are automatically transferred to the other graphs. An elaborate example of this can be found in chapter 9. This interactive exploration of data facilitates the discovery of new insights.

Figure 2.18 shows the average score per country for questions. Not only does this indicate a high correlation between the answers, but it's easy to see that when you select several points on a subplot, the points will correspond to similar points on the other graphs. In this case the selected points on the left graph correspond to points on the middle and right graphs, although they correspond better in the middle and right graphs.

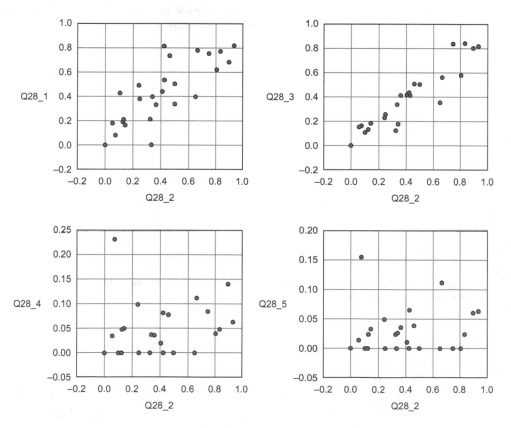

Figure 2.16 Drawing multiple plots together can help you understand the structure of your data over multiple variables.

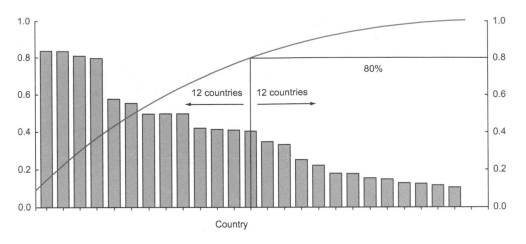

Figure 2.17 A Pareto diagram is a combination of the values and a cumulative distribution. It's easy to see from this diagram that the first 50% of the countries contain slightly less than 80% of the total amount. If this graph represented customer buying power and we sell expensive products, we probably don't need to spend our marketing budget in every country; we could start with the first 50%.

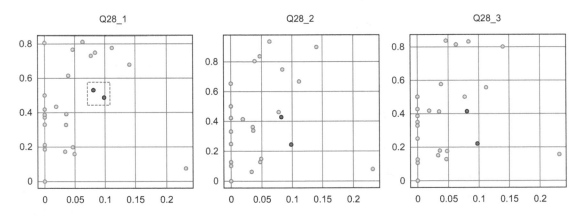

Figure 2.18 Link and brush allows you to select observations in one plot and highlight the same observations in the other plots.

Two other important graphs are the histogram shown in figure 2.19 and the boxplot shown in figure 2.20.

In a histogram a variable is cut into discrete categories and the number of occurrences in each category are summed up and shown in the graph. The boxplot, on the other hand, doesn't show how many observations are present but does offer an impression of the distribution within categories. It can show the maximum, minimum, median, and other characterizing measures at the same time.

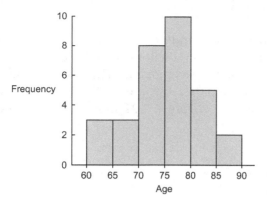

Figure 2.19   Example histogram: the number of people in the age-groups of 5-year intervals

The techniques we described in this phase are mainly visual, but in practice they're certainly not limited to visualization techniques. Tabulation, clustering, and other modeling techniques can also be a part of exploratory analysis. Even building simple models can be a part of this step.

Now that you've finished the data exploration phase and you've gained a good grasp of your data, it's time to move on to the next phase: building models.

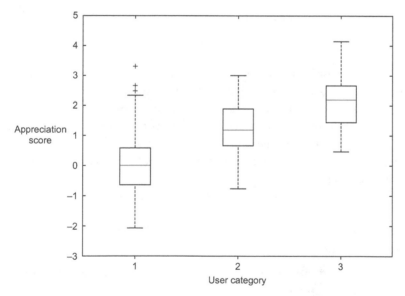

Figure 2.20   Example boxplot: each user category has a distribution of the appreciation each has for a certain picture on a photography website.

2.6   *Step 5: Build the models*

With clean data in place and a good understanding of the content, you're ready to build models with the goal of making better predictions, classifying objects, or gaining an understanding of the system that you're modeling. This phase is much more focused than the exploratory analysis step, because you know what you're looking for and what you want the outcome to be. Figure 2.21 shows the components of model building.

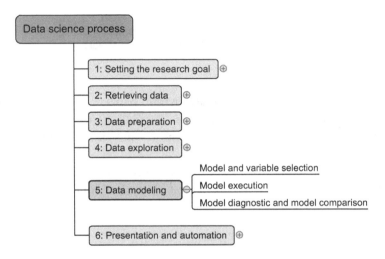

Figure 2.21   Step 5: Data modeling

The techniques you'll use now are borrowed from the field of machine learning, data mining, and/or statistics. In this chapter we only explore the tip of the iceberg of existing techniques, while chapter 3 introduces them properly. It's beyond the scope of this book to give you more than a conceptual introduction, but it's enough to get you started; 20% of the techniques will help you in 80% of the cases because techniques overlap in what they try to accomplish. They often achieve their goals in similar but slightly different ways.

Building a model is an iterative process. The way you build your model depends on whether you go with classic statistics or the somewhat more recent machine learning school, and the type of technique you want to use. Either way, most models consist of the following main steps:

1   Selection of a modeling technique and variables to enter in the model
2   Execution of the model
3   Diagnosis and model comparison

2.6.1   *Model and variable selection*

You'll need to select the variables you want to include in your model and a modeling technique. Your findings from the exploratory analysis should already give a fair idea of what variables will help you construct a good model. Many modeling techniques are available, and choosing the right model for a problem requires judgment on your part. You'll need to consider model performance and whether your project meets all the requirements to use your model, as well as other factors:

- Must the model be moved to a production environment and, if so, would it be easy to implement?
- How difficult is the maintenance on the model: how long will it remain relevant if left untouched?
- Does the model need to be easy to explain?

When the thinking is done, it's time for action.

2.6.2 Model execution

Once you've chosen a model you'll need to implement it in code.

> **REMARK** This is the first time we'll go into actual Python code execution so make sure you have a virtual env up and running. Knowing how to set this up is required knowledge, but if it's your first time, check out appendix D.
>
> All code from this chapter can be downloaded from https://www.manning .com/books/introducing-data-science. This chapter comes with an ipython (.ipynb) notebook and Python (.py) file.

Luckily, most programming languages, such as Python, already have libraries such as StatsModels or Scikit-learn. These packages use several of the most popular techniques. Coding a model is a nontrivial task in most cases, so having these libraries available can speed up the process. As you can see in the following code, it's fairly easy to use linear regression (figure 2.22) with StatsModels or Scikit-learn. Doing this yourself would require much more effort even for the simple techniques. The following listing shows the execution of a linear prediction model.

Listing 2.1 Executing a linear prediction model on semi-random data

```
import statsmodels.api as sm              Imports required
import numpy as np                        Python modules.
predictors = np.random.random(1000).reshape(500,2)
target = predictors.dot(np.array([0.4, 0.6])) + np.random.random(500)
lmRegModel = sm.OLS(target,predictors)
result = lmRegModel.fit()
result.summary()
```

Shows model fit statistics.

Fits linear regression on data.

Creates random data for predictors (x-values) and semi-random data for the target (y-values) of the model. We use predictors as input to create the target so we infer a correlation here.

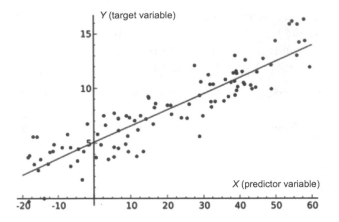

Figure 2.22 Linear regression tries to fit a line while minimizing the distance to each point

Okay, we cheated here, quite heavily so. We created predictor values that are meant to predict how the target variables behave. For a linear regression, a "linear relation" between each x (predictor) and the y (target) variable is assumed, as shown in figure 2.22.

We, however, created the target variable, based on the predictor by adding a bit of randomness. It shouldn't come as a surprise that this gives us a well-fitting model. The `results.summary()` outputs the table in figure 2.23. Mind you, the exact outcome depends on the random variables you got.

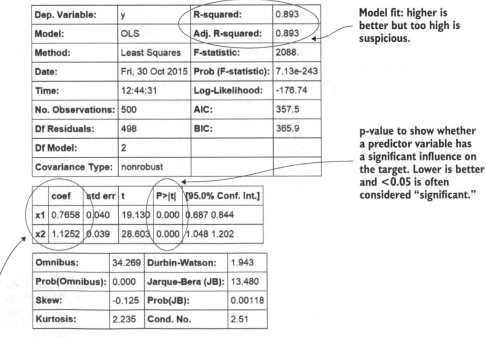

Dep. Variable:	y	R-squared:	0.893
Model:	OLS	Adj. R-squared:	0.893
Method:	Least Squares	F-statistic:	2088.
Date:	Fri, 30 Oct 2015	Prob (F-statistic):	7.13e-243
Time:	12:44:31	Log-Likelihood:	-176.74
No. Observations:	500	AIC:	357.5
Df Residuals:	498	BIC:	365.9
Df Model:	2		
Covariance Type:	nonrobust		

Model fit: higher is better but too high is suspicious.

	coef	std err	t	P>|t|	[95.0% Conf. Int.]
x1	0.7658	0.040	19.130	0.000	0.687 0.844
x2	1.1252	0.039	28.603	0.000	1.048 1.202

p-value to show whether a predictor variable has a significant influence on the target. Lower is better and <0.05 is often considered "significant."

Omnibus:	34.269	Durbin-Watson:	1.943
Prob(Omnibus):	0.000	Jarque-Bera (JB):	13.480
Skew:	-0.125	Prob(JB):	0.00118
Kurtosis:	2.235	Cond. No.	2.51

Linear equation coefficients.
$y = 0.7658x1 + 1.1252x2.$

Figure 2.23   Linear regression model information output

Let's ignore most of the output we got here and focus on the most important parts:

- *Model fit*—For this the R-squared or adjusted R-squared is used. This measure is an indication of the amount of variation in the data that gets captured by the model. The difference between the adjusted R-squared and the R-squared is minimal here because the adjusted one is the normal one + a penalty for model complexity. A model gets complex when many variables (or features) are introduced. You don't need a complex model if a simple model is available, so the adjusted R-squared punishes you for overcomplicating. At any rate, 0.893 is high, and it should be because we cheated. Rules of thumb exist, but for models in businesses, models above 0.85 are often considered good. If you want to win a competition you need in the high 90s. For research however, often very low model fits (<0.2 even) are found. What's more important there is the influence of the introduced predictor variables.

- *Predictor variables have a coefficient*—For a linear model this is easy to interpret. In our example if you add "1" to x1, it will change y by "0.7658". It's easy to see how finding a good predictor can be your route to a Nobel Prize even though your model as a whole is rubbish. If, for instance, you determine that a certain gene is significant as a cause for cancer, this is important knowledge, even if that gene in itself doesn't determine whether a person will get cancer. The example here is classification, not regression, but the point remains the same: detecting influences is more important in scientific studies than perfectly fitting models (not to mention more realistic). But when do we know a gene has that impact? This is called significance.

- *Predictor significance*—Coefficients are great, but sometimes not enough evidence exists to show that the influence is there. This is what the p-value is about. A long explanation about type 1 and type 2 mistakes is possible here but the short explanations would be: if the p-value is lower than 0.05, the variable is considered significant for most people. In truth, this is an arbitrary number. It means there's a 5% chance the predictor doesn't have any influence. Do you accept this 5% chance to be wrong? That's up to you. Several people introduced the extremely significant ($p<0.01$) and marginally significant thresholds ($p<0.1$).

Linear regression works if you want to predict a value, but what if you want to classify something? Then you go to classification models, the best known among them being k-nearest neighbors.

As shown in figure 2.24, k-nearest neighbors looks at labeled points nearby an unlabeled point and, based on this, makes a prediction of what the label should be.

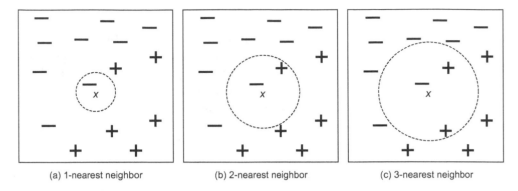

(a) 1-nearest neighbor   (b) 2-nearest neighbor   (c) 3-nearest neighbor

Figure 2.24   K-nearest neighbor techniques look at the k-nearest point to make a prediction.

Let's try it in Python code using the Scikit learn library, as in this next listing.

Listing 2.2   Executing k-nearest neighbor classification on semi-random data

```
from sklearn import neighbors                ←——  Imports modules.
predictors = np.random.random(1000).reshape(500,2)
target = np.around(predictors.dot(np.array([0.4, 0.6])) +
        np.random.random(500))
clf = neighbors.KNeighborsClassifier(n_neighbors=10)
knn = clf.fit(predictors,target)
knn.score(predictors, target)
```

Creates random predictor data and semi-random target data based on predictor data.

Fits 10-nearest neighbors model.

Gets model fit score: what percent of the classification was correct?

As before, we construct random correlated data and surprise, surprise we get 85% of cases correctly classified. If we want to look in depth, we need to score the model. Don't let `knn.score()` fool you; it returns the model accuracy, but by "scoring a model" we often mean applying it on data to make a prediction.

```
prediction = knn.predict(predictors)
```

Now we can use the prediction and compare it to the real thing using a confusion matrix.

```
metrics.confusion_matrix(target,prediction)
```

We get a 3-by-3 matrix as shown in figure 2.25.

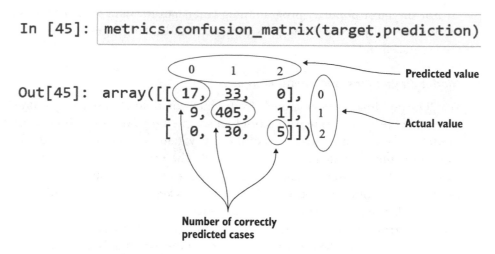

```
In [45]: metrics.confusion_matrix(target,prediction)
```

Out[45]: array([[17, 33,   0],
                 [9, 405,   1],
                 [0, 30,   5]])

Predicted value

Actual value

Number of correctly
predicted cases

Figure 2.25   Confusion matrix: it shows how many cases were correctly classified and incorrectly classified by comparing the prediction with the real values. Remark: the classes (0,1,2) were added in the figure for clarification.

The confusion matrix shows we have correctly predicted 17+405+5 cases, so that's good. But is it really a surprise? No, for the following reasons:

- For one, the classifier had but three options; marking the difference with last time np.around() will round the data to its nearest integer. In this case that's either 0, 1, or 2. With only 3 options, you can't do much worse than 33% correct on 500 guesses, even for a real random distribution like flipping a coin.

- Second, we cheated again, correlating the response variable with the predictors. Because of the way we did this, we get most observations being a "1". By guessing "1" for every case we'd already have a similar result.

- We compared the prediction with the real values, true, but we never predicted based on fresh data. The prediction was done using the same data as the data used to build the model. This is all fine and dandy to make yourself feel good, but it gives you no indication of whether your model will work when it encounters truly new data. For this we need a holdout sample, as will be discussed in the next section.

Don't be fooled. Typing this code won't work miracles by itself. It might take a while to get the modeling part and all its parameters right.

To be honest, only a handful of techniques have industry-ready implementations in Python. But it's fairly easy to use models that are available in R within Python with the help of the RPy library. RPy provides an interface from Python to R. *R* is a free software environment, widely used for statistical computing. If you haven't already, it's worth at least a look, because in 2014 it was still one of the most popular (if not the most popular) programming languages for data science. For more information, see http://www.kdnuggets.com/polls/2014/languages-analytics-data-mining-data-science.html.

2.6.3   *Model diagnostics and model comparison*

You'll be building multiple models from which you then choose the best one based on multiple criteria. Working with a holdout sample helps you pick the best-performing model. A holdout sample is a part of the data you leave out of the model building so it can be used to evaluate the model afterward. The principle here is simple: the model should work on unseen data. You use only a fraction of your data to estimate the model and the other part, the holdout sample, is kept out of the equation. The model is then unleashed on the unseen data and error measures are calculated to evaluate it. Multiple error measures are available, and in figure 2.26 we show the general idea on comparing models. The error measure used in the example is the mean square error.

$$MSE = \frac{1}{n} \sum_{i=1}^{n} (\hat{Y_i} - Y_i)^2$$

Figure 2.26   Formula for mean square error

Mean square error is a simple measure: check for every prediction how far it was from the truth, square this error, and add up the error of every prediction.

Figure 2.27 compares the performance of two models to predict the order size from the price. The first model is *size = 3 * price* and the second model is *size = 10*. To

	n	Size	Price	Predicted model 1	Predicted model 2	Error model 1	Error model 2
	1	10	3				
	2	15	5				
	3	18	6				
	4	14	5				
	...	...					
80% train	800	9	3				
	801	12	4	12	10	0	2
	802	13	4	12	10	1	3
	...						
	999	21	7	21	10	0	11
20% test	1000	10	4	12	10	−2	0
					Total	5861	110225

Figure 2.27   A holdout sample helps you compare models and ensures that you can generalize results to data that the model has not yet seen.

estimate the models, we use 800 randomly chosen observations out of 1,000 (or 80%), without showing the other 20% of data to the model. Once the model is trained, we predict the values for the other 20% of the variables based on those for which we already know the true value, and calculate the model error with an error measure. Then we choose the model with the lowest error. In this example we chose model 1 because it has the lowest total error.

Many models make strong assumptions, such as independence of the inputs, and you have to verify that these assumptions are indeed met. This is called *model diagnostics*.

This section gave a short introduction to the steps required to build a valid model. Once you have a working model you're ready to go to the last step.

2.7   *Step 6: Presenting findings and building applications on top of them*

After you've successfully analyzed the data and built a well-performing model, you're ready to present your findings to the world (figure 2.28). This is an exciting part; all your hours of hard work have paid off and you can explain what you found to the stakeholders.

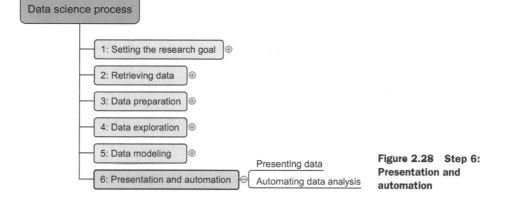

Figure 2.28   Step 6: Presentation and automation

Sometimes people get so excited about your work that you'll need to repeat it over and over again because they value the predictions of your models or the insights that you produced. For this reason, you need to automate your models. This doesn't always mean that you have to redo all of your analysis all the time. Sometimes it's sufficient that you implement only the model scoring; other times you might build an application that automatically updates reports, Excel spreadsheets, or PowerPoint presentations. The last stage of the data science process is where your *soft skills* will be most useful, and yes, they're extremely important. In fact, we recommend you find dedicated books and other information on the subject and work through them, because why bother doing all this tough work if nobody listens to what you have to say?

If you've done this right, you now have a working model and satisfied stakeholders, so we can conclude this chapter here.

2.8   *Summary*

In this chapter you learned the data science process consists of six steps:

- *Setting the research goal*—Defining the what, the why, and the how of your project in a project charter.
- *Retrieving data*—Finding and getting access to data needed in your project. This data is either found within the company or retrieved from a third party.
- *Data preparation*—Checking and remediating data errors, enriching the data with data from other data sources, and transforming it into a suitable format for your models.
- *Data exploration*—Diving deeper into your data using descriptive statistics and visual techniques.
- *Data modeling*—Using machine learning and statistical techniques to achieve your project goal.
- *Presentation and automation*—Presenting your results to the stakeholders and industrializing your analysis process for repetitive reuse and integration with other tools.

Machine learning 3

This chapter covers

- Understanding why data scientists use machine learning
- Identifying the most important Python libraries for machine learning
- Discussing the process for model building
- Using machine learning techniques
- Gaining hands-on experience with machine learning

Do you know how computers learn to protect you from malicious persons? Computers filter out more than 60% of your emails and can learn to do an even better job at protecting you over time.

Can you explicitly teach a computer to recognize persons in a picture? It's possible but impractical to encode all the possible ways to recognize a person, but you'll soon see that the possibilities are nearly endless. To succeed, you'll need to add a new skill to your toolkit, *machine learning*, which is the topic of this chapter.

3.1   *What is machine learning and why should you care about it?*

> "Machine learning is a field of study that gives computers the ability to learn without being explicitly programmed."
>
> —Arthur Samuel, 1959[1]

The definition of machine learning coined by Arthur Samuel is often quoted and is genius in its broadness, but it leaves you with the question of how the computer learns. To achieve machine learning, experts develop general-purpose algorithms that can be used on large classes of learning problems. When you want to solve a *specific task* you only need to feed the algorithm more *specific data*. In a way, you're programming by example. In most cases a computer will use data as its source of information and compare its output to a desired output and then correct for it. The more data or "experience" the computer gets, the better it becomes at its designated job, like a human does.

When machine learning is seen as a process, the following definition is insightful:

> "Machine learning is the process by which a computer can work more accurately as it collects and learns from the data it is given."
>
> —Mike Roberts[2]

For example, as a user writes more text messages on a phone, the phone learns more about the messages' common vocabulary and can predict (autocomplete) their words faster and more accurately.

In the broader field of science, machine learning is a subfield of artificial intelligence and is closely related to applied mathematics and statistics. All this might sound a bit abstract, but machine learning has many applications in everyday life.

3.1.1   *Applications for machine learning in data science*

Regression and *classification* are of primary importance to a data scientist. To achieve these goals, one of the main tools a data scientist uses is machine learning. The uses for regression and automatic classification are wide ranging, such as the following:

- Finding oil fields, gold mines, or archeological sites based on existing sites (classification and regression)
- Finding place names or persons in text (classification)
- Identifying people based on pictures or voice recordings (classification)
- Recognizing birds based on their whistle (classification)

---

[1] Although the following paper is often cited as the source of this quote, it's not present in a 1967 reprint of that paper. The authors were unable to verify or find the exact source of this quote. See Arthur L. Samuel, "Some Studies in Machine Learning Using the Game of Checkers," *IBM Journal of Research and Development* 3, no. 3 (1959):210–229.

[2] Mike Roberts is the technical editor of this book. Thank you, Mike.

- Identifying profitable customers (regression and classification)
- Proactively identifying car parts that are likely to fail (regression)
- Identifying tumors and diseases (classification)
- Predicting the amount of money a person will spend on product X (regression)
- Predicting the number of eruptions of a volcano in a period (regression)
- Predicting your company's yearly revenue (regression)
- Predicting which team will win the Champions League in soccer (classification)

Occasionally data scientists build a *model* (an abstraction of reality) that provides insight to the underlying processes of a phenomenon. When the goal of a model isn't prediction but interpretation, it's called *root cause analysis*. Here are a few examples:

- Understanding and optimizing a business process, such as determining which products add value to a product line
- Discovering what causes diabetes
- Determining the causes of traffic jams

This list of machine learning applications can only be seen as an appetizer because it's ubiquitous within data science. Regression and classification are two important techniques, but the repertoire and the applications don't end, with clustering as one other example of a valuable technique. Machine learning techniques can be used throughout the data science process, as we'll discuss in the next section.

3.1.2   *Where machine learning is used in the data science process*

Although machine learning is mainly linked to the data-modeling step of the data science process, it can be used at almost every step. To refresh your memory from previous chapters, the data science process is shown in figure 3.1.

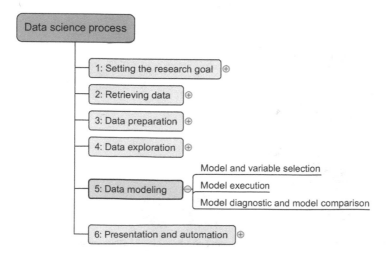

Figure 3.1   The data science process

The data modeling phase can't start until you have qualitative raw data you can understand. But prior to that, the *data preparation* phase can benefit from the use of machine learning. An example would be cleansing a list of text strings; machine learning can group similar strings together so it becomes easier to correct spelling errors.

Machine learning is also useful when *exploring data*. Algorithms can root out underlying patterns in the data where they'd be difficult to find with only charts.

Given that machine learning is useful throughout the data science process, it shouldn't come as a surprise that a considerable number of Python libraries were developed to make your life a bit easier.

3.1.3 *Python tools used in machine learning*

Python has an overwhelming number of packages that can be used in a machine learning setting. The Python machine learning ecosystem can be divided into three main types of packages, as shown in figure 3.2.

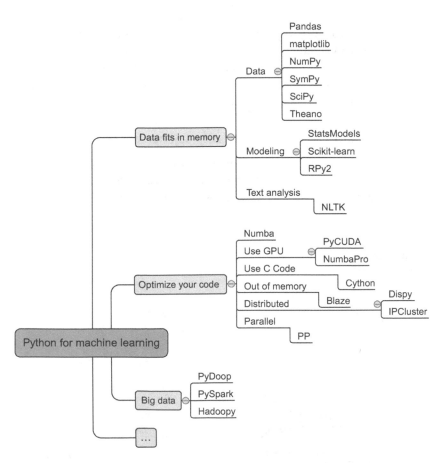

Figure 3.2   Overview of Python packages used during the machine-learning phase

The first type of package shown in figure 3.2 is mainly used in simple tasks and when data fits into memory. The second type is used to optimize your code when you've finished prototyping and run into speed or memory issues. The third type is specific to using Python with big data technologies.

PACKAGES FOR WORKING WITH DATA IN MEMORY

When prototyping, the following packages can get you started by providing advanced functionalities with a few lines of code:

- *SciPy* is a library that integrates fundamental packages often used in scientific computing such as NumPy, matplotlib, Pandas, and SymPy.
- *NumPy* gives you access to powerful array functions and linear algebra functions.
- *Matplotlib* is a popular 2D plotting package with some 3D functionality.
- *Pandas* is a high-performance, but easy-to-use, data-wrangling package. It introduces dataframes to Python, a type of in-memory data table. It's a concept that should sound familiar to regular users of R.
- *SymPy* is a package used for symbolic mathematics and computer algebra.
- *StatsModels* is a package for statistical methods and algorithms.
- *Scikit-learn* is a library filled with machine learning algorithms.
- *RPy2* allows you to call R functions from within Python. R is a popular open source statistics program.
- *NLTK* (Natural Language Toolkit) is a Python toolkit with a focus on text analytics.

These libraries are good to get started with, but once you make the decision to run a certain Python program at frequent intervals, performance comes into play.

OPTIMIZING OPERATIONS

Once your application moves into production, the libraries listed here can help you deliver the speed you need. Sometimes this involves connecting to big data infrastructures such as Hadoop and Spark.

- *Numba and NumbaPro*—These use just-in-time compilation to speed up applications written directly in Python and a few annotations. NumbaPro also allows you to use the power of your graphics processor unit (GPU).
- *PyCUDA*—This allows you to write code that will be executed on the GPU instead of your CPU and is therefore ideal for calculation-heavy applications. It works best with problems that lend themselves to being parallelized and need little input compared to the number of required computing cycles. An example is studying the robustness of your predictions by calculating thousands of different outcomes based on a single start state.
- *Cython, or C for Python*—This brings the C programming language to Python. C is a lower-level language, so the code is closer to what the computer eventually uses (bytecode). The closer code is to bits and bytes, the faster it executes. A computer is also faster when it knows the type of a variable (called *static typing*). Python wasn't designed to do this, and Cython helps you to overcome this shortfall.

- *Blaze*—Blaze gives you data structures that can be bigger than your computer's main memory, enabling you to work with large data sets.
- *Dispy and IPCluster*—These packages allow you to write code that can be distributed over a cluster of computers.
- *PP*—Python is executed as a single process by default. With the help of PP you can parallelize computations on a single machine or over clusters.
- *Pydoop and Hadoopy*—These connect Python to Hadoop, a common big data framework.
- *PySpark*—This connects Python and Spark, an in-memory big data framework.

Now that you've seen an overview of the available libraries, let's look at the modeling process itself.

3.2   *The modeling process*

The modeling phase consists of four steps:

1 Feature engineering and model selection
2 Training the model
3 Model validation and selection
4 Applying the trained model to unseen data

Before you find a good model, you'll probably iterate among the first three steps.

The last step isn't always present because sometimes the goal isn't prediction but explanation (root cause analysis). For instance, you might want to find out the causes of species' extinctions but not necessarily predict which one is next in line to leave our planet.

It's possible to *chain* or *combine* multiple techniques. When you chain multiple models, the output of the first model becomes an input for the second model. When you combine multiple models, you train them independently and combine their results. This last technique is also known as *ensemble learning*.

A model consists of constructs of information called *features* or *predictors* and a *target* or *response variable*. Your model's goal is to predict the target variable, for example, tomorrow's high temperature. The variables that help you do this and are (usually) known to you are the features or predictor variables such as today's temperature, cloud movements, current wind speed, and so on. The best models are those that accurately represent reality, preferably while staying concise and interpretable. To achieve this, feature engineering is the most important and arguably most interesting part of modeling. For example, an important feature in a model that tried to explain the extinction of large land animals in the last 60,000 years in Australia turned out to be the population number and spread of humans.

3.2.1   *Engineering features and selecting a model*

With engineering features, you must come up with and create possible predictors for the model. This is one of the most important steps in the process because a model recombines these features to achieve its predictions. Often you may need to consult an expert or the appropriate literature to come up with meaningful features.

Certain features are the variables you get from a data set, as is the case with the provided data sets in our exercises and in most school exercises. In practice you'll need to find the features yourself, which may be scattered among different data sets. In several projects we had to bring together more than 20 different data sources before we had the raw data we required. Often you'll need to apply a transformation to an input before it becomes a good predictor or to combine multiple inputs. An example of combining multiple inputs would be *interaction variables*: the impact of either single variable is low, but if both are present their impact becomes immense. This is especially true in chemical and medical environments. For example, although vinegar and bleach are fairly harmless common household products by themselves, mixing them results in poisonous chlorine gas, a gas that killed thousands during World War I.

In medicine, clinical pharmacy is a discipline dedicated to researching the effect of the interaction of medicines. This is an important job, and it doesn't even have to involve two medicines to produce potentially dangerous results. For example, mixing an antifungal medicine such as Sporanox with grapefruit has serious side effects.

Sometimes you have to use modeling techniques to derive features: the output of a model becomes part of another model. This isn't uncommon, especially in text mining. Documents can first be annotated to classify the content into categories, or you can count the number of geographic places or persons in the text. This counting is often more difficult than it sounds; models are first applied to recognize certain words as a person or a place. All this new information is then poured into the model you want to build. One of the biggest mistakes in model construction is the *availability bias*: your features are only the ones that you could easily get your hands on and your model consequently represents this one-sided "truth." Models suffering from availability bias often fail when they're validated because it becomes clear that they're not a valid representation of the truth.

In World War II, after bombing runs on German territory, many of the English planes came back with bullet holes in the wings, around the nose, and near the tail of the plane. Almost none of them had bullet holes in the cockpit, tail rudder, or engine block, so engineering decided extra armor plating should be added to the wings. This looked like a sound idea until a mathematician by the name of Abraham Wald explained the obviousness of their mistake: they only took into account the planes that returned. The bullet holes on the wings were actually the least of their concern, because at least a plane with this kind of damage could make it back home for repairs. Plane fortification was hence increased on the spots that were unscathed on returning planes. The initial reasoning suffered from availability bias: the engineers ignored an important part of the data because it was harder to obtain. In this case they were lucky, because the reasoning could be reversed to get the intended result without getting the data from the crashed planes.

When the initial features are created, a model can be *trained* to the data.

3.2.2 Training your model

With the right predictors in place and a modeling technique in mind, you can progress to model training. In this phase you present to your model data from which it can learn.

The most common modeling techniques have industry-ready implementations in almost every programming language, including Python. These enable you to train your models by executing a few lines of code. For more state-of-the art data science techniques, you'll probably end up doing heavy mathematical calculations and implementing them with modern computer science techniques.

Once a model is trained, it's time to test whether it can be extrapolated to reality: model validation.

3.2.3 Validating a model

Data science has many modeling techniques, and the question is which one is the right one to use. A good model has two properties: it has good predictive power and it generalizes well to data it hasn't seen. To achieve this you define an error measure (how wrong the model is) and a validation strategy.

Two common *error measures* in machine learning are the *classification error rate* for classification problems and the *mean squared error* for regression problems. The classification error rate is the percentage of observations in the test data set that your model mislabeled; lower is better. The mean squared error measures how big the average error of your prediction is. Squaring the average error has two consequences: you can't cancel out a wrong prediction in one direction with a faulty prediction in the other direction. For example, overestimating future turnover for next month by 5,000 doesn't cancel out underestimating it by 5,000 for the following month. As a second consequence of squaring, bigger errors get even more weight than they otherwise would. Small errors remain small or can even shrink (if <1), whereas big errors are enlarged and will definitely draw your attention.

Many *validation strategies* exist, including the following common ones:

- *Dividing your data into a training set with X% of the observations and keeping the rest as a holdout data set* (a data set that's never used for model creation)—This is the most common technique.
- *K-folds cross validation*—This strategy divides the data set into k parts and uses each part one time as a test data set while using the others as a training data set. This has the advantage that you use all the data available in the data set.
- *Leave-1 out*—This approach is the same as k-folds but with k=1. You always leave one observation out and train on the rest of the data. This is used only on small data sets, so it's more valuable to people evaluating laboratory experiments than to big data analysts.

Another popular term in machine learning is *regularization*. When applying regularization, you incur a penalty for every extra variable used to construct the model. With *L1 regularization* you ask for a model with as few predictors as possible. This is important for the model's robustness: simple solutions tend to hold true in more situations. *L2 regularization* aims to keep the variance between the coefficients of the predictors as small as possible. Overlapping variance between predictors makes it hard to make out the actual impact of each predictor. Keeping their variance from overlapping will increase interpretability. To keep it simple: regularization is mainly used to stop a model from using too many features and thus prevent over-fitting.

Validation is extremely important because it determines whether your model works in real-life conditions. To put it bluntly, it's whether your model is worth a dime. Even so, every now and then people send in papers to respected scientific journals (and sometimes even succeed at publishing them) with faulty validation. The result of this is they get rejected or need to retract the paper because everything is wrong. Situations like this are bad for your mental health so always keep this in mind: test your models on data the constructed model has never seen and make sure this data is a true representation of what it would encounter when applied on fresh observations by other people. For classification models, instruments like the confusion matrix (introduced in chapter 2 but thoroughly explained later in this chapter) are golden; embrace them.

Once you've constructed a good model, you can (optionally) use it to predict the future.

3.2.4   *Predicting new observations*

If you've implemented the first three steps successfully, you now have a performant model that generalizes to unseen data. The process of applying your model to new data is called model scoring. In fact, model scoring is something you implicitly did during validation, only now you don't know the correct outcome. By now you should trust your model enough to use it for real.

Model scoring involves two steps. First, you prepare a data set that has features exactly as defined by your model. This boils down to repeating the data preparation you did in step one of the modeling process but for a new data set. Then you apply the model on this new data set, and this results in a prediction.

Now let's look at the different types of machine learning techniques: a different problem requires a different approach.

3.3   *Types of machine learning*

Broadly speaking, we can divide the different approaches to machine learning by the amount of human effort that's required to coordinate them and how they use *labeled data*—data with a category or a real-value number assigned to it that represents the outcome of previous observations.

- *Supervised learning* techniques attempt to discern results and learn by trying to find patterns in a labeled data set. Human interaction is required to label the data.
- *Unsupervised learning* techniques don't rely on labeled data and attempt to find patterns in a data set without human interaction.

- *Semi-supervised learning* techniques need labeled data, and therefore human interaction, to find patterns in the data set, but they can still progress toward a result and learn even if passed unlabeled data as well.

In this section, we'll look at all three approaches, see what tasks each is more appropriate for, and use one or two of the Python libraries mentioned earlier to give you a feel for the code and solve a task. In each of these examples, we'll work with a downloadable data set that has already been cleaned, so we'll skip straight to the data modeling step of the data science process, as discussed earlier in this chapter.

3.3.1 Supervised learning

As stated before, supervised learning is a learning technique that can only be applied on labeled data. An example implementation of this would be discerning digits from images. Let's dive into a case study on number recognition.

CASE STUDY: DISCERNING DIGITS FROM IMAGES

One of the many common approaches on the web to stopping computers from hacking into user accounts is the Captcha check—a picture of text and numbers that the human user must decipher and enter into a form field before sending the form back to the web server. Something like figure 3.3 should look familiar.

Figure 3.3   A simple Captcha control can be used to prevent automated spam being sent through an online web form.

With the help of the *Naïve Bayes classifier*, a simple yet powerful algorithm to categorize observations into classes that's explained in more detail in the sidebar, you can recognize digits from textual images. These images aren't unlike the Captcha checks many websites have in place to make sure you're not a computer trying to hack into the user accounts. Let's see how hard it is to let a computer recognize images of numbers.

Our research goal is to let a computer recognize images of numbers (step one of the data science process).

The data we'll be working on is the MNIST data set, which is often used in the data science literature for teaching and benchmarking.

Introducing Naïve Bayes classifiers in the context of a spam filter

Not every email you receive has honest intentions. Your inbox can contain unsolicited commercial or bulk emails, a.k.a. spam. Not only is spam annoying, it's often used in scams and as a carrier for viruses. Kaspersky[3] estimates that more than 60% of the emails in the world are spam. To protect users from spam, most email clients run a program in the background that classifies emails as either spam or safe.

A popular technique in spam filtering is employing a classifier that uses the words inside the mail as predictors. It outputs the chance that a specific email is spam given the words it's composed of (in mathematical terms, P(spam | words)). To reach this conclusion it uses three calculations:

- P(spam)—The average rate of spam without knowledge of the words. According to Kaspersky, an email is spam 60% of the time.
- P(words)—How often this word combination is used regardless of spam.
- P(words | spam)—How often these words are seen when a training mail was labeled as spam.

To determine the chance that a new email is spam, you'd use the following formula:

P(spam|words) = P(spam)P(words|spam) / P(words)

This is an application of the rule P(B|A) = P(B) P(A|B) / P(A), which is known as Bayes's rule and which lends its name to this classifier. The "naïve" part comes from the classifier's assumption that the presence of one feature doesn't tell you anything about another feature (feature independence, also called absence of multicollinearity). In reality, features are often related, especially in text. For example the word "buy" will often be followed by "now." Despite the unrealistic assumption, the naïve classifier works surprisingly well in practice.

With the bit of theory in the sidebar, you're ready to perform the modeling itself. Make sure to run all the upcoming code in the same scope because each piece requires the one before it. An IPython file can be downloaded for this chapter from the Manning download page of this book.

The MNIST images can be found in the data sets package of Scikit-learn and are already normalized for you (all scaled to the same size: 64x64 pixels), so we won't need much data preparation (step three of the data science process). But let's first fetch our data as step two of the data science process, with the following listing.

Listing 3.1   Step 2 of the data science process: fetching the digital image data

```
from sklearn.datasets import load_digits        ◁──┐   Imports digits
import pylab as pl                                       database.
digits = load_digits()    ◁──┐   Loads digits.
```

---

[3] Kaspersky 2014 Quarterly Spam Statistics Report, http://usa.kaspersky.com/internet-security-center/threats/spam-statistics-report-q1-2014#.VVym9blViko.

Working with images isn't much different from working with other data sets. In the case of a gray image, you put a value in every matrix entry that depicts the gray value to be shown. The following code demonstrates this process and is step four of the data science process: data exploration.

Listing 3.2   Step 4 of the data science process: using Scikit-learn

```
pl.gray()
pl.matshow(digits.images[0])        Shows first       Turns image into
pl.show()                           images.           gray-scale values.
digits.images[0]

                                    Shows the
                                    corresponding matrix.
```

Figure 3.4 shows how a blurry "0" image translates into a data matrix.

Figure 3.4 shows the actual code output, but perhaps figure 3.5 can clarify this slightly, because it shows how each element in the vector is a piece of the image.

Easy so far, isn't it? There is, naturally, a little more work to do. The Naïve Bayes classifier is expecting a list of values, but `pl.matshow()` returns a two-dimensional array (a matrix) reflecting the shape of the image. To flatten it into a list, we need to call `reshape()` on `digits.images`. The net result will be a one-dimensional array that looks something like this:

```
array([[ 0., 0., 5., 13., 9., 1., 0., 0., 0., 0., 13., 15., 10., 15., 5., 0.,
0., 3., 15., 2., 0., 11., 8., 0., 0., 4., 12., 0., 0., 8., 8., 0.,
0., 5., 8., 0., 0., 9., 8., 0., 0., 4., 11., 0., 1., 12., 7., 0.,
0., 2., 14., 5., 10., 12., 0., 0., 0., 0., 6., 13., 10., 0., 0., 0.]])
```

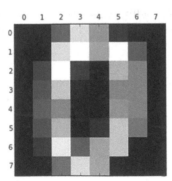

```
rray([[ 0.,  0.,  5., 13.,  9.,  1.,  0.,  0.],
      [ 0.,  0., 13., 15., 10., 15.,  5.,  0.],
      [ 0.,  3., 15.,  2.,  0., 11.,  8.,  0.],
      [ 0.,  4., 12.,  0.,  0.,  8.,  8.,  0.],
      [ 0.,  5.,  8.,  0.,  0.,  9.,  8.,  0.],
      [ 0.,  4., 11.,  0.,  1., 12.,  7.,  0.],
      [ 0.,  2., 14.,  5., 10., 12.,  0.,  0.],
      [ 0.,  0.,  6., 13., 10.,  0.,  0.,  0.]])
```

Figure 3.4   Blurry grayscale representation of the number 0 with its corresponding matrix. The higher the number, the closer it is to white; the lower the number, the closer it is to black.

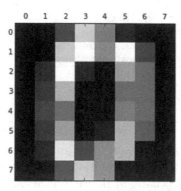 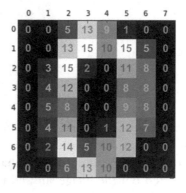

Figure 3.5 We'll turn an image into something usable by the Naïve Bayes classifier by getting the grayscale value for each of its pixels (shown on the right) and putting those values in a list.

The previous code snippet shows the matrix of figure 3.5 flattened (the number of dimensions was reduced from two to one) to a Python list. From this point on, it's a standard classification problem, which brings us to step five of the data science process: model building.

Now that we have a way to pass the contents of an image into the classifier, we need to pass it a training data set so it can start learning how to predict the numbers in the images. We mentioned earlier that Scikit-learn contains a subset of the MNIST database (1,800 images), so we'll use that. Each image is also labeled with the number it actually shows. This will build a probabilistic model in memory of the most likely digit shown in an image given its grayscale values.

Once the program has gone through the training set and built the model, we can then pass it the test set of data to see how well it has learned to interpret the images using the model.

The following listing shows how to implement these steps in code.

Listing 3.3 Image data classification problem on images of digits

```
from sklearn.cross_validation import train_test_split
from sklearn.naive_bayes import GaussianNB
from sklearn.metrics import confusion_matrix
import pylab as plt


y = digits.target


n_samples = len(digits.images)
X= digits.images.reshape((n_samples, -1))

print X

X_train, X_test, y_train, y_test = train_test_split(X, y, random_state=0)
```

Step 1: Select target variable.

Step 2: Prepare data. Reshape adapts the matrix form. This method could, for instance, turn a 10x10 matrix into 100 vectors.

Step 3: Split into test set and training set.

Step 5:
Fit data.

```
gnb = GaussianNB()
fit = gnb.fit(X_train,y_train)
predicted = fit.predict(X_test)
confusion_matrix(y_test, predicted)
```

Step 4: Select a Naïve Bayes classifier; use a Gaussian distribution to estimate probability.

Step 7: Create confusion matrix.

Step 6: Predict data for unseen data.

The end result of this code is called a *confusion matrix*, such as the one shown in figure 3.6. Returned as a two-dimensional array, it shows how often the number predicted was the correct number on the main diagonal and also in the matrix entry (i,j), where j was predicted but the image showed i. Looking at figure 3.6 we can see that the model predicted the number 2 correctly 17 times (at coordinates 3,3), but also that the model predicted the number 8 15 times when it was actually the number 2 in the image (at 9,3).

```
array([[37,  0,  0,  0,  0,  0,  0,  0,  0,  0],
       [ 0, 39,  0,  0,  0,  0,  1,  0,  3,  0],
       [ 0,  9, 17,  3,  0,  0,  0,  0, 15,  0],
       [ 0,  0,  0, 38,  0,  0,  0,  2,  5,  0],
       [ 0,  1,  0,  0, 27,  0,  2,  8,  0,  0],
       [ 0,  1,  0,  1,  0, 43,  0,  3,  0,  0],
       [ 0,  0,  0,  0,  0,  0, 52,  0,  0,  0],
       [ 0,  0,  0,  0,  1,  0,  0, 47,  0,  0],
       [ 0,  5,  0,  1,  0,  1,  0,  4, 37,  0],
       [ 0,  2,  0,  7,  1,  0,  0,  3,  7, 27]]])
```

Figure 3.6 Confusion matrix produced by predicting what number is depicted by a blurry image

Confusion matrices

A confusion matrix is a matrix showing how wrongly (or correctly) a model predicted, how much it got "confused." In its simplest form it will be a 2x2 table for models that try to classify observations as being A or B. Let's say we have a classification model that predicts whether somebody will buy our newest product: deep-fried cherry pudding. We can either predict: "Yes, this person will buy" or "No, this customer won't buy." Once we make our prediction for 100 people we can compare this to their actual behavior, showing us how many times we got it right. An example is shown in table 3.1.

Table 3.1 Confusion matrix example

Confusion matrix	Predicted "Person will buy"	Predicted "Person will not buy"
Person **bought** the deep-fried cherry pudding	35 (true positive)	10 (false negative)
Person **didn't buy** the deep-fried cherry pudding	15 (false positive)	40 (true negative)

The model was correct in (35+40) 75 cases and incorrect in (15+10) 25 cases, resulting in a (75 correct/100 total observations) 75% accuracy.

All the correctly classified observations are added up on the diagonal (35+40) while everything else (15+10) is incorrectly classified. When the model only predicts two classes (binary), our correct guesses are two groups: true positives (predicted to buy and did so) and true negatives (predicted they wouldn't buy and they didn't). Our incorrect guesses are divided into two groups: false positives (predicted they would buy but they didn't) and false negatives (predicted not to buy but they did). The matrix is useful to see where the model is having the most problems. In this case we tend to be overconfident in our product and classify customers as future buyers too easily (false positive).

From the confusion matrix, we can deduce that for most images the predictions are quite accurate. In a good model you'd expect the sum of the numbers on the main diagonal of the matrix (also known as the matrix *trace*) to be very high compared to the sum of all matrix entries, indicating that the predictions were correct for the most part.

Let's assume we want to show off our results in a more easily understandable way or we want to inspect several of the images and the predictions our program has made: we can use the following code to display one next to the other. Then we can see where the program has gone wrong and needs a little more training. If we're satisfied with the results, the model building ends here and we arrive at step six: presenting the results.

Listing 3.4 Inspecting predictions vs actual numbers

Adds an extra subplot on a 6x3 plot grid. This code could be simplified as: plt.subplot (3, 2 ,index) but this looks visually more appealing.

Stores number image matrix and its prediction (as a number) together in array.

Loops through first 7 images.

```python
images_and_predictions = list(zip(digits.images, fit.predict(X)))
for index, (image, prediction) in enumerate(images_and_predictions[:6]):
    plt.subplot(6, 3 ,index + 5)
    plt.axis('off')
    plt.imshow(image, cmap=plt.cm.gray_r, interpolation='nearest')
    plt.title('Prediction: %i' % prediction)
plt.show()
```

Doesn't show an axis.

Shows image in grayscale.

Shows the full plot that is now populated with 6 subplots.

Shows the predicted value as the title to the shown image.

Figure 3.7 shows how all predictions seem to be correct except for the digit number 2, which it labels as 8. We should forgive this mistake as this 2 does share visual similarities

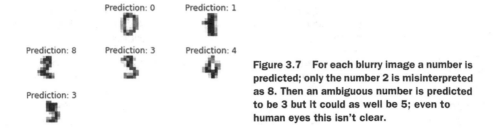

Figure 3.7   **For each blurry image a number is predicted; only the number 2 is misinterpreted as 8. Then an ambiguous number is predicted to be 3 but it could as well be 5; even to human eyes this isn't clear.**

with 8. The bottom left number is ambiguous, even to humans; is it a 5 or a 3? It's debatable, but the algorithm thinks it's a 3.

By discerning which images were misinterpreted, we can train the model further by labeling them with the correct number they display and feeding them back into the model as a new training set (step 5 of the data science process). This will make the model more accurate, so the cycle of learn, predict, correct continues and the predictions become more accurate. This is a controlled data set we're using for the example. All the examples are the same size and they are all in 16 shades of gray. Expand that up to the variable size images of variable length strings of variable shades of alphanumeric characters shown in the Captcha control, and you can appreciate why a model accurate enough to predict any Captcha image doesn't exist yet.

In this supervised learning example, it's apparent that without the labels associated with each image telling the program what number that image shows, a model cannot be built and predictions cannot be made. By contrast, an unsupervised learning approach doesn't need its data to be labeled and can be used to give structure to an unstructured data set.

3.3.2 *Unsupervised learning*

It's generally true that most large data sets don't have labels on their data, so unless you sort through it all and give it labels, the supervised learning approach to data won't work. Instead, we must take the approach that will work with this data because

- We can study the *distribution of the data* and infer truths about the data in different parts of the distribution.
- We can study the *structure and values in the data* and infer new, more meaningful data and structure from it.

Many techniques exist for each of these *unsupervised learning* approaches. However, in the real world you're always working toward the research goal defined in the first phase of the data science process, so you may need to combine or try different techniques before either a data set can be labeled, enabling supervised learning techniques, perhaps, or even the goal itself is achieved.

DISCERNING A SIMPLIFIED LATENT STRUCTURE FROM YOUR DATA

Not everything can be measured. When you meet someone for the first time you might try to guess whether they like you based on their behavior and how they respond. But what if they've had a bad day up until now? Maybe their cat got run over or they're still down from attending a funeral the week before? The point is that certain variables can be immediately available while others can only be inferred and are therefore missing from your data set. The first type of variables are known as *observable variables* and the second type are known as *latent variables.* In our example, the emotional state of your new friend is a latent variable. It definitely influences their judgment of you but its value isn't clear.

Deriving or inferring latent variables and their values based on the actual contents of a data set is a valuable skill to have because

- Latent variables can substitute for several existing variables already in the data set.
- By reducing the number of variables in the data set, the data set becomes more manageable, any further algorithms run on it work faster, and predictions may become more accurate.
- Because latent variables are designed or targeted toward the defined research goal, you lose little key information by using them.

If we can reduce a data set from 14 observable variables per line to 5 or 6 latent variables, for example, we have a better chance of reaching our research goal because of the data set's simplified structure. As you'll see from the example below, it's not a case of reducing the existing data set to as few latent variables as possible. You'll need to find the sweet spot where the number of latent variables derived returns the most value. Let's put this into practice with a small case study.

CASE STUDY: FINDING LATENT VARIABLES IN A WINE QUALITY DATA SET

In this short case study, you'll use a technique known as Principal Component Analysis (PCA) to find latent variables in a data set that describes the quality of wine. Then you'll compare how well a set of latent variables works in predicting the quality of wine against the original observable set. You'll learn

1. How to identify and derive those latent variables.
2. How to analyze where the sweet spot is—how many new variables return the most utility—by generating and interpreting a *scree plot* generated by PCA. (We'll look at scree plots in a moment.)

Let's look at the main components of this example.

- *Data set*—The University of California, Irvine (UCI) has an online repository of 325 data sets for machine learning exercises at http://archive.ics.uci.edu/ml/. We'll use the Wine Quality Data Set for red wines created by P. Cortez, A. Cerdeira, F. Almeida, T. Matos, and J. Reis[4]. It's 1,600 lines long and has 11 variables per line, as shown in table 3.2.

---

[4] You can find full details of the Wine Quality Data Set at https://archive.ics.uci.edu/ml/datasets/Wine+Quality.

Table 3.2　The first three rows of the Red Wine Quality Data Set

Fixed acidity	Volatile acidity	Citric acid	Residual sugar	Chlorides	Free sulfur dioxide	Total sulfur dioxide	Density	pH	Sulfates	Alcohol	Quality
7.4	0.7	0	1.9	0.076	11	34	0.9978	3.51	0.56	9.4	5
7.8	0.88	0	2.6	0.098	25	67	0.9968	3.2	0.68	9.8	5
7.8	0.76	0.04	2.3	0.092	15	54	0.997	3.26	0.65	9.8	5

- *Principal Component Analysis*—A technique to find the latent variables in your data set while retaining as much information as possible.
- *Scikit-learn*—We use this library because it already implements PCA for us and is a way to generate the scree plot.

Part one of the data science process is to set our research goal: We want to explain the subjective "wine quality" feedback using the different wine properties.

Our first job then is to download the data set (step two: acquiring data), as shown in the following listing, and prepare it for analysis (step three: data preparation). Then we can run the PCA algorithm and view the results to look at our options.

Listing 3.5　Data acquisition and variable standardization

X is a matrix of predictor variables. These variables are wine properties such as density and alcohol presence.

```
import pandas as pd
from sklearn import preprocessing
from sklearn.decomposition import PCA
import pylab as plt
from sklearn import preprocessing

url = http://archive.ics.uci.edu/ml/machine-learning-databases/wine-quality/
    winequality-red.csv
data = pd.read_csv(url, sep= ";")
X = data[[u'fixed acidity', u'volatile acidity', u'citric acid',
    u'residual sugar', u'chlorides', u'free sulfur dioxide',
    u'total sulfur dioxide', u'density', u'pH', u'sulphates',
    u'alcohol']]
y = data.quality
X= preprocessing.StandardScaler().fit(X).transform(X)
```

Downloads location of wine-quality data set.

Reads in the CSV data. It's separated by a semi-colon.

y is a vector and represents the dependent variable (target variable). y is the perceived wine quality.

When standardizing data, the following formula is applied to every data point: $z = (x-\mu)/\sigma$, where z is the new observation value, x the old one, μ is the mean, and σ the standard deviation. The PCA of a data matrix is easier to interpret when the columns have first been centered by their means.

With the initial data preparation behind you, you can execute the PCA. The resulting scree plot (which will be explained shortly) is shown in figure 3.8. Because PCA is an explorative technique, we now arrive at step four of the data science process: data exploration, as shown in the following listing.

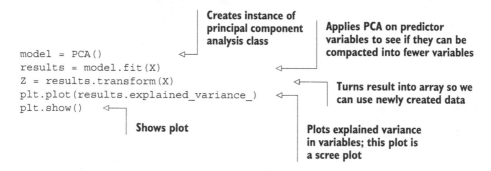

Listing 3.6 Executing the principal component analysis

```
model = PCA()
results = model.fit(X)
Z = results.transform(X)
plt.plot(results.explained_variance_)
plt.show()
```

Creates instance of principal component analysis class

Applies PCA on predictor variables to see if they can be compacted into fewer variables

Turns result into array so we can use newly created data

Shows plot

Plots explained variance in variables; this plot is a scree plot

Now let's look at the scree plot in figure 3.8.

The plot generated from the wine data set is shown in figure 3.8. What you hope to see is an elbow or hockey stick shape in the plot. This indicates that a few variables can represent the majority of the information in the data set while the rest only add a little more. In our plot, PCA tells us that reducing the set down to one variable can capture approximately 28% of the total information in the set (the plot is zero-based, so variable

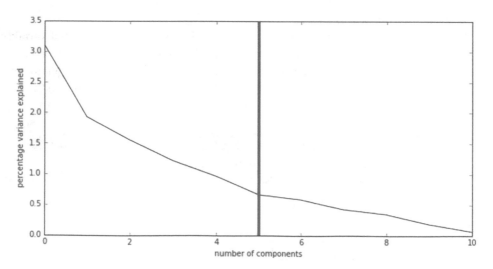

Figure 3.8 PCA scree plot showing the marginal amount of information of every new variable PCA can create. The first variables explain approximately 28% of the variance in the data, the second variable accounts for another 17%, the third approximately 15%, and so on.

one is at position zero on the x axis), two variables will capture approximately 17% more or 45% total, and so on. Table 3.3 shows you the full read-out.

Table 3.3 The findings of the PCA

Number of variables	Extra information captured	Total data captured
1	28%	28%
2	17%	45%
3	14%	59%
4	10%	69%
5	8%	77%
6	7%	84%
7	5%	89%
8 - 11	...	100%

An elbow shape in the plot suggests that five variables can hold most of the information found inside the data. You could argue for a cut-off at six or seven variables instead, but we're going to opt for a simpler data set versus one with less variance in data against the original data set.

At this point, we could go ahead and see if the original data set recoded with five latent variables is good enough to predict the quality of the wine accurately, but before we do, we'll see how we might identify what they represent.

INTERPRETING THE NEW VARIABLES

With the initial decision made to reduce the data set from 11 original variables to 5 latent variables, we can check to see whether it's possible to interpret or name them based on their relationships with the originals. Actual names are easier to work with than codes such as lv1, lv2, and so on. We can add the line of code in the following listing to generate a table that shows how the two sets of variables correlate.

Listing 3.7 **Showing PCA components in a Pandas data frame**

```
pd.DataFrame(results.components_, columns=list(
    [u'fixed acidity', u'volatile acidity', u'citric acid', u'residual sugar',
        u'chlorides', u'free sulfur dioxide',  u'total sulfur dioxide', u'density',
        u'pH', u'sulphates',  u'alcohol']))
```

The rows in the resulting table (table 3.4) show the mathematical correlation. Or, in English, the first latent variable lv1, which captures approximately 28% of the total information in the set, has the following formula.

```
Lv1 = (fixed acidity * 0.489314) + (volatile acidity * -0.238584) + … +
(alcohol * -0.113232)
```

Table 3.4 How PCA calculates the 11 original variables' correlation with 5 latent variables

	Fixed acidity	Volatile acidity	Citric acid	Residual sugar	Chlorides	Free sulfur dioxide	Total sulfur dioxide	Density	pH	Sulphates	Alcohol
0	0.489314	-0.238584	0.463632	0.146107	0.212247	-0.036158	0.023575	0.395353	-0.438520	0.242921	-0.113232
1	-0.110503	0.274930	-0.151791	0.272080	0.148052	0.513567	0.569487	0.233575	0.006711	-0.037554	-0.386181
2	0.123302	0.449963	-0.238247	-0.101283	0.092614	-0.428793	-0.322415	0.338871	-0.057697	-0.279786	-0.471673
3	-0.229617	0.078960	-0.079418	-0.372793	0.666195	-0.043538	-0.034577	-0.174500	-0.003788	0.550872	-0.122181
4	0.082614	-0.218735	0.058573	-0.732144	-0.246501	0.159152	0.222465	-0.157077	-0.267530	-0.225962	-0.350681

Giving a useable name to each new variable is a bit trickier and would probably require consultation with an actual wine expert for accuracy. However, as we don't have a wine expert on hand, we'll call them the following (table 3.5).

Table 3.5 Interpretation of the wine quality PCA-created variables

Latent variable	Possible interpretation
0	Persistent acidity
1	Sulfides
2	Volatile acidity
3	Chlorides
4	Lack of residual sugar

We can now recode the original data set with only the five latent variables. Doing this is data preparation again, so we revisit step three of the data science process: data preparation. As mentioned in chapter 2, the data science process is a recursive one and this is especially true between step three: data preparation and step 4: data exploration.

Table 3.6 shows the first three rows with this done.

Table 3.6 The first three rows of the Red Wine Quality Data Set recoded in five latent variables

	Persistent acidity	Sulfides	Volatile acidity	Chlorides	Lack of residual sugar
0	-1.619530	0.450950	**1.774454**	0.043740	-0.067014
1	-0.799170	1.856553	0.911690	0.548066	0.018392
2	**2.357673**	-0.269976	-0.243489	-0.928450	**1.499149**

Already we can see high values for wine 0 in volatile acidity, while wine 2 is particularly high in persistent acidity. Don't sound like good wines at all!

COMPARING THE ACCURACY OF THE ORIGINAL DATA SET WITH LATENT VARIABLES

Now that we've decided our data set should be recoded into 5 latent variables rather than the 11 originals, it's time to see how well the new data set works for predicting the quality of wine when compared to the original. We'll use the Naïve Bayes Classifier algorithm we saw in the previous example for supervised learning to help.

Let's start by seeing how well the original 11 variables could predict the wine quality scores. The following listing presents the code to do this.

Listing 3.8 Wine score prediction before principal component analysis

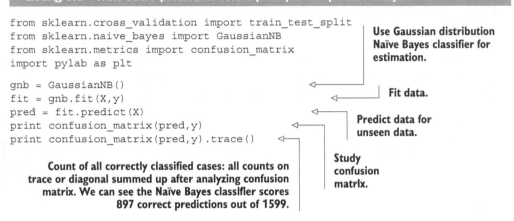

```
from sklearn.cross_validation import train_test_split
from sklearn.naive_bayes import GaussianNB
from sklearn.metrics import confusion_matrix
import pylab as plt

gnb = GaussianNB()
fit = gnb.fit(X,y)
pred = fit.predict(X)
print confusion_matrix(pred,y)
print confusion_matrix(pred,y).trace()
```

Use Gaussian distribution Naïve Bayes classifier for estimation.

Fit data.

Predict data for unseen data.

Study confusion matrix.

Count of all correctly classified cases: all counts on trace or diagonal summed up after analyzing confusion matrix. We can see the Naïve Bayes classifier scores 897 correct predictions out of 1599.

Now we'll run the same prediction test, but starting with only 1 latent variable instead of the original 11. Then we'll add another, see how it did, add another, and so on to see how the predictive performance improves. The following listing shows how this is done.

Listing 3.9 Wine score prediction with increasing number of principal components

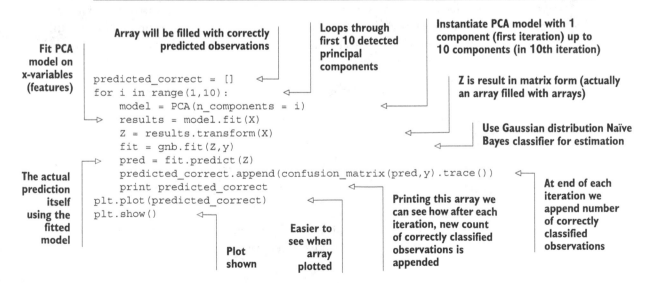

Fit PCA model on x-variables (features)

Array will be filled with correctly predicted observations

Loops through first 10 detected principal components

Instantiate PCA model with 1 component (first iteration) up to 10 components (in 10th iteration)

```
predicted_correct = []
for i in range(1,10):
    model = PCA(n_components = i)
    results = model.fit(X)
    Z = results.transform(X)
    fit = gnb.fit(Z,y)
    pred = fit.predict(Z)
    predicted_correct.append(confusion_matrix(pred,y).trace())
    print predicted_correct
plt.plot(predicted_correct)
plt.show()
```

Z is result in matrix form (actually an array filled with arrays)

Use Gaussian distribution Naïve Bayes classifier for estimation

The actual prediction itself using the fitted model

Plot shown

Easier to see when array plotted

Printing this array we can see how after each iteration, new count of correctly classified observations is appended

At end of each iteration we append number of correctly classified observations

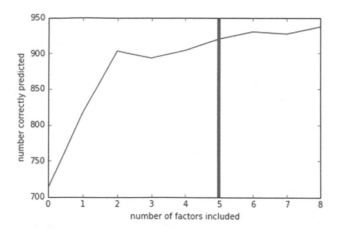

number correctly predicted (y-axis)

number of factors included (x-axis)

Figure 3.9 The results plot shows that adding more latent variables to a model (x-axis) greatly increases predictive power (y-axis) up to a point but then tails off. The gain in predictive power from adding variables wears off eventually.

The resulting plot is shown in figure 3.9.

The plot in figure 3.9 shows that with only 3 latent variables, the classifier does a better job of predicting wine quality than with the original 11. Also, adding more latent variables beyond 5 doesn't add as much predictive power as the first 5. This shows our choice of cutting off at 5 variables was a good one, as we'd hoped.

We looked at how to group similar variables, but it's also possible to group observations.

GROUPING SIMILAR OBSERVATIONS TO GAIN INSIGHT FROM THE DISTRIBUTION OF YOUR DATA

Suppose for a moment you're building a website that recommends films to users based on preferences they've entered and films they've watched. The chances are high that if they watch many horror movies they're likely to want to know about new horror movies and not so much about new teen romance films. By grouping together users who've watched more or less the same films and set more or less the same preferences, you can gain a good deal of insight into what else they might like to have recommended.

The general technique we're describing here is known as *clustering*. In this process, we attempt to divide our data set into observation subsets, or *clusters*, wherein observations should be similar to those in the same cluster but differ greatly from the observations in other clusters. Figure 3.10 gives you a visual idea of what clustering aims to achieve. The circles in the top left of the figure are clearly close to each other while being farther away from the others. The same is true of the crosses in the top right.

Scikit-learn implements several common algorithms for clustering data in its `sklearn.cluster` module, including the k-means algorithm, affinity propagation, and spectral clustering. Each has a use case or two for which it's more suited,[5] although

---

[5] You can find a comparison of all the clustering algorithms in Scikit-learn at http://scikit-learn.org/stable/modules/clustering.html.

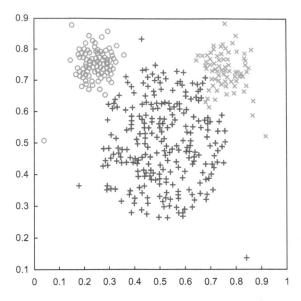

Figure 3.10 **The goal of clustering is to divide a data set into "sufficiently distinct" subsets. In this plot for instance, the observations have been divided into three clusters.**

k-means is a good general-purpose algorithm with which to get started. However, like all the clustering algorithms, you need to specify the number of desired clusters in advance, which necessarily results in a process of trial and error before reaching a decent conclusion. It also presupposes that all the data required for analysis is available already. What if it wasn't?

Let's look at the actual case of clustering irises (the flower) by their properties (sepal length and width, petal length and width, and so on). In this example we'll use the k-means algorithm. It's a good algorithm to get an impression of the data but it's sensitive to start values, so you can end up with a different cluster every time you run the algorithm unless you manually define the start values by specifying a seed (constant for the start value generator). If you need to detect a hierarchy, you're better off using an algorithm from the class of hierarchical clustering techniques.

One other disadvantage is the need to specify the number of desired clusters in advance. This often results in a process of trial and error before coming to a satisfying conclusion.

Executing the code is fairly simple. It follows the same structure as all the other analyses except you don't have to pass a target variable. It's up to the algorithm to learn interesting patterns. The following listing uses an iris data set to see if the algorithm can group the different types of irises.

Listing 3.10   Iris classification example

Print first 5 observations of data frame to screen; now we can clearly see 4 variables: sepal length, sepal width, petal length, and petal width.

Add another variable called "cluster" to data frame. This indicates the cluster membership of every flower in data set.

Load in iris (flowers) data of Scikit-learn.

Fit model to data. All variables are considered independent variables; unsupervised learning has no target variable (y).

```python
import sklearn
from sklearn import cluster
import pandas as pd

data = sklearn.datasets.load_iris()
X = pd.DataFrame(data.data, columns = list(data.feature_names))
print X[:5]
model    = cluster.KMeans(n_clusters=3, random_state=25)
results = model.fit(X)
X["cluster"] = results.predict(X)
X["target"] = data.target
X["c"] = "lookatmeIamimportant"
print X[:5]
classification_result = X[["cluster",
       "target","c"]].groupby(["cluster","target"]).agg("count")
print(classification_result)
```

Transform iris data into Pandas data frame.

Let's finally add a target variable (y) to the data frame.

Adding a variable c is just a little trick we use to do a count later. The value here is arbitrary because we need a column to count the rows.

Three parts to this code. First we select the cluster, target, and c columns. Then we group by the cluster and target columns. Finally, we aggregate the row of the group with a simple count aggregation.

Initialize a k-means cluster model with 3 clusters. The random_state is a random seed; if you don't put it in, the seed will also be random. We opt for 3 clusters because we saw in the last listing this might be a good compromise between complexity and performance.

The matrix this classification result represents gives us an indication of whether our clustering was successful. For cluster 0, we're spot on. On clusters 1 and 2 there has been a slight mix-up, but in total we only get 16 (14+2) misclassifications out of 150.

Figure 3.11 shows the output of the iris classification.

This figure shows that even without using a label you'd find clusters that are similar to the official iris classification with a result of 134 (50+48+36) correct classifications out of 150.

You don't always need to choose between supervised and unsupervised; sometimes combining them is an option.

cluster	target	c
0	0	50
1	1	48
1	2	14
2	1	2
2	2	36

Figure 3.11 Output of the iris classification

3.4   *Semi-supervised learning*

It shouldn't surprise you to learn that while we'd like all our data to be labeled so we can use the more powerful supervised machine learning techniques, in reality we often start with only minimally labeled data, if it's labeled at all. We can use our unsupervised machine learning techniques to analyze what we have and perhaps add labels to the data set, but it will be prohibitively costly to label it all. Our goal then is to train our predictor models with as little labeled data as possible. This is where semi-supervised learning techniques come in—hybrids of the two approaches we've already seen.

Take for example the plot in figure 3.12. In this case, the data has only two labeled observations; normally this is too few to make valid predictions.

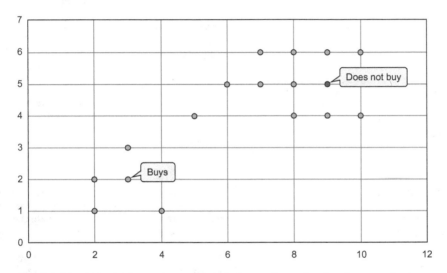

Figure 3.12   This plot has only two labeled observations—too few for supervised observations, but enough to start with an unsupervised or semi-supervised approach.

A common semi-supervised learning technique is *label propagation*. In this technique, you start with a labeled data set and give the same label to similar data points. This is similar to running a clustering algorithm over the data set and labeling each cluster based on the labels they contain. If we were to apply this approach to the data set in figure 3.12, we might end up with something like figure 3.13.

One special approach to semi-supervised learning worth mentioning here is *active learning*. In active learning the program points out the observations it wants to see labeled for its next round of learning based on some criteria you have specified. For example, you might set it to try and label the observations the algorithm is least certain about, or you might use multiple models to make a prediction and select the points where the models disagree the most.

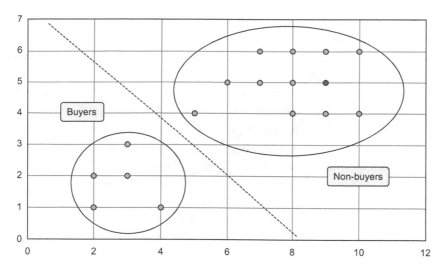

Figure 3.13 The previous figure shows that the data has only two labeled observations, far too few for supervised learning. This figure shows how you can exploit the structure of the underlying data set to learn better classifiers than from the labeled data only. The data is split into two clusters by the clustering technique; we only have two labeled values, but if we're bold we can assume others within that cluster have that same label (buyer or non-buyer), as depicted here. This technique isn't flawless; it's better to get the actual labels if you can.

With the basics of machine learning at your disposal, the next chapter discusses using machine learning within the constraints of a single computer. This tends to be challenging when the data set is too big to load entirely into memory.

3.5 Summary

In this chapter, you learned that

- Data scientists rely heavily on techniques from statistics and machine learning to perform their modeling. A good number of real-life applications exist for machine learning, from classifying bird whistling to predicting volcanic eruptions.
- The modeling process consists of four phases:
  1 *Feature engineering, data preparation, and model parameterization*—We define the input parameters and variables for our model.
  2 *Model training*—The model is fed with data and it learns the patterns hidden in the data.
  3 *Model selection and validation*—A model can perform well or poorly; based on its performance we select the model that makes the most sense.
  4 *Model scoring*—When our model can be trusted, it's unleashed on new data. If we did our job well, it will provide us with extra insights or give us a good prediction of what the future holds.

- The two big types of machine learning techniques
  1. *Supervised*—Learning that requires labeled data.
  2. *Unsupervised*—Learning that doesn't require labeled data but is usually less accurate or reliable than supervised learning.
- Semi-supervised learning is in between those techniques and is used when only a small portion of the data is labeled.
- Two case studies demonstrated supervised and unsupervised learning, respectively:
  1. Our first case study made use of a Naïve Bayes classifier to classify images of numbers as the number they represent. We also took a look at the confusion matrix as a means to determining how well our classification model is doing.
  2. Our case study on unsupervised techniques showed how we could use principal component analysis to reduce the input variables for further model building while maintaining most of the information.

Handling large data on
a single computer

4

This chapter covers

- Working with large data sets on a single computer
- Working with Python libraries suitable for larger data sets
- Understanding the importance of choosing correct algorithms and data structures
- Understanding how you can adapt algorithms to work inside databases

What if you had so much data that it seems to outgrow you, and your techniques no longer seem to suffice? What do you do, surrender or adapt?

Luckily you chose to adapt, because you're still reading. This chapter introduces you to techniques and tools to handle larger data sets that are still manageable by a single computer if you adopt the right techniques.

This chapter gives you the tools to perform the classifications and regressions when the data no longer fits into the RAM (random access memory) of your computer, whereas chapter 3 focused on in-memory data sets. Chapter 5 will go a step further and teach you how to deal with data sets that require multiple computers to be processed. When we refer to *large data* in this chapter we mean data that causes problems to work with in terms of memory or speed but can still be handled by a single computer.

We start this chapter with an overview of the problems you face when handling large data sets. Then we offer three types of solutions to overcome these problems: adapt your algorithms, choose the right data structures, and pick the right tools. Data scientists aren't the only ones who have to deal with large data volumes, so you can apply general best practices to tackle the large data problem. Finally, we apply this knowledge to two case studies. The first case shows you how to detect malicious URLs, and the second case demonstrates how to build a recommender engine inside a database.

4.1   *The problems you face when handling large data*

A large volume of data poses new challenges, such as overloaded memory and algorithms that never stop running. It forces you to adapt and expand your repertoire of techniques. But even when you can perform your analysis, you should take care of issues such as I/O (input/output) and CPU starvation, because these can cause speed issues. Figure 4.1 shows a mind map that will gradually unfold as we go through the steps: problems, solutions, and tips.

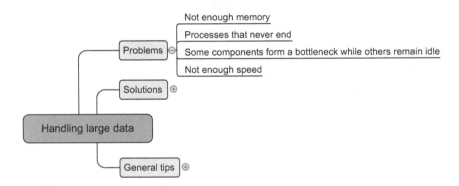

Figure 4.1   Overview of problems encountered when working with more data than can fit in memory

A computer only has a limited amount of RAM. When you try to squeeze more data into this memory than actually fits, the OS will start swapping out memory blocks to disks, which is far less efficient than having it all in memory. But only a few algorithms are designed to handle large data sets; most of them load the whole data set into memory at once, which causes the out-of-memory error. Other algorithms need to hold multiple copies of the data in memory or store intermediate results. All of these aggravate the problem.

Even when you cure the memory issues, you may need to deal with another limited resource: *time.* Although a computer may think you live for millions of years, in reality you won't (unless you go into cryostasis until your PC is done). Certain algorithms don't take time into account; they'll keep running forever. Other algorithms can't end in a reasonable amount of time when they need to process only a few megabytes of data.

A third thing you'll observe when dealing with large data sets is that components of your computer can start to form a bottleneck while leaving other systems idle. Although this isn't as severe as a never-ending algorithm or out-of-memory errors, it still incurs a serious cost. Think of the cost savings in terms of person days and computing infrastructure for CPU starvation. Certain programs don't feed data fast enough to the processor because they have to read data from the hard drive, which is one of the slowest components on a computer. This has been addressed with the introduction of solid state drives (SSD), but SSDs are still much more expensive than the slower and more widespread hard disk drive (HDD) technology.

4.2   General techniques for handling large volumes of data

Never-ending algorithms, out-of-memory errors, and speed issues are the most common challenges you face when working with large data. In this section, we'll investigate solutions to overcome or alleviate these problems.

The solutions can be divided into three categories: using the correct algorithms, choosing the right data structure, and using the right tools (figure 4.2).

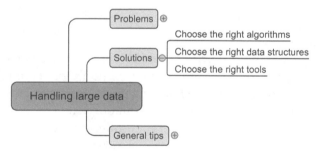

Figure 4.2   Overview of solutions for handling large data sets

No clear one-to-one mapping exists between the problems and solutions because many solutions address both lack of memory and computational performance. For instance, data set compression will help you solve memory issues because the data set becomes smaller. But this also affects computation speed with a shift from the slow hard disk to the fast CPU. Contrary to RAM (random access memory), the hard disc will store everything even after the power goes down, but writing to disc costs more time than changing information in the fleeting RAM. When constantly changing the information, RAM is thus preferable over the (more durable) hard disc. With an unpacked data set, numerous read and write operations (I/O) are occurring, but the CPU remains largely idle, whereas with the compressed data set the CPU gets its fair share of the workload. Keep this in mind while we explore a few solutions.

4.2.1 *Choosing the right algorithm*

Choosing the right algorithm can solve more problems than adding more or better hardware. An algorithm that's well suited for handling large data doesn't need to load the entire data set into memory to make predictions. Ideally, the algorithm also supports parallelized calculations. In this section we'll dig into three types of algorithms that can do that: *online algorithms*, *block algorithms*, and *MapReduce algorithms*, as shown in figure 4.3.

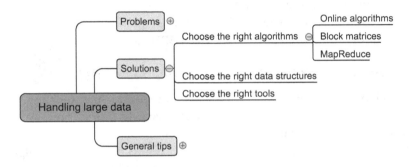

Figure 4.3 Overview of techniques to adapt algorithms to large data sets

ONLINE LEARNING ALGORITHMS

Several, but not all, machine learning algorithms can be trained using one observation at a time instead of taking all the data into memory. Upon the arrival of a new data point, the model is trained and the observation can be forgotten; its effect is now incorporated into the model's parameters. For example, a model used to predict the weather can use different parameters (like atmospheric pressure or temperature) in different regions. When the data from one region is loaded into the algorithm, it forgets about this raw data and moves on to the next region. This "use and forget" way of working is the perfect solution for the memory problem as a single observation is unlikely to ever be big enough to fill up all the memory of a modern-day computer.

Listing 4.1 shows how to apply this principle to a perceptron with online learning. A *perceptron* is one of the least complex machine learning algorithms used for binary classification (0 or 1); for instance, will the customer buy or not?

Listing 4.1   Training a perceptron by observation

The learning rate of an algorithm is the adjustment it makes every time a new observation comes in. If this is high, the model will adjust quickly to new observations but might "overshoot" and never get precise. An oversimplified example: the optimal (and unknown) weight for an x-variable = 0.75. Current estimation is 0.4 with a learning rate of 0.5; the adjustment = 0.5 (learning rate) * 1(size of error) * 1 (value of x) = 0.5. 0.4 (current weight) + 0.5 (adjustment) = 0.9 (new weight), instead of 0.75. The adjustment was too big to get the correct result.

```python
import numpy as np
class perceptron():
    def __init__(self, X,y, threshold = 0.5,
learning_rate = 0.1, max_epochs = 10):
        self.threshold = threshold
        self.learning_rate = learning_rate
        self.X = X
        self.y = y
        self.max_epochs = max_epochs
```

Sets up perceptron class.

The __init__ method of any Python class is always run when creating an instance of the class. Several default values are set here.

The threshold is an arbitrary cutoff between 0 and 1 to decide whether the prediction becomes a 0 or a 1. Often it's 0.5, right in the middle, but it depends on the use case.

X and y variables are assigned to the class.

One epoch is one run through all the data. We allow for a maximum of 10 runs until we stop the perceptron.

Each observation will end up with a weight. The initialize function sets these weights for each incoming observation. We allow for 2 options: all weights start at 0 or they are assigned a small (between 0 and 0.05) random weight.

```python
    def initialize(self, init_type = 'zeros'):
        if init_type == 'random':
            self.weights = np.random.rand(len(self.X[0])) * 0.05
        if init_type == 'zeros':
            self.weights = np.zeros(len(self.X[0]))
```

We start at the first epoch.

The training function.

True is always true, so technically this is a never-ending loop, but we build in several stop (break) conditions.

```python
    def train(self):
        epoch = 0
        while True:
            error_count = 0
            epoch += 1
            for (X,y) in zip(self.X, self.y):
                error_count += self.train_observation(X,y,error_count)
```

Adds one to the current number of epochs.

Initiates the number of encountered errors at 0 for each epoch. This is important; if an epoch ends without errors, the algorithm converged and we're done.

We loop through the data and feed it to the train observation function, one observation at a time.

```
        if error_count == 0:
            print "training successful"
            break
        if epoch >= self.max_epochs:
            print "reached maximum epochs, no perfect prediction"
            break
```

If we reach the maximum number of allowed runs, we stop looking for a solution.

If by the end of the epoch we don't have an error, the training was successful.

The real value (y) is either 0 or 1; the prediction is also 0 or 1. If it's wrong we get an error of either 1 or -1.

The train observation function is run for every observation and will adjust the weights using the formula explained earlier.

```
    def train_observation(self,X,y, error_count):
        result = np.dot(X, self.weights) > self.threshold
        error = y - result
```

A prediction is made for this observation. Because it's binary, this will be either 0 or 1.

In case we have a wrong prediction (an error), we need to adjust the model.

Adds 1 to the error count.

For every predictor variable in the input vector (X), we'll adjust its weight.

```
        if error != 0:
            error_count += 1
            for index, value in enumerate(X):
                self.weights[index] += self.learning_rate * error * value
        return error_count
```

We return the error count because we need to evaluate it at the end of the epoch.

Adjusts the weight for every predictor variable using the learning rate, the error, and the actual value of the predictor variable.

The predict class.

```
    def predict(self, X):
        return int(np.dot(X, self.weights) > self.threshold)
```

The values of the predictor values are multiplied by their respective weights (this multiplication is done by np.dot). Then the outcome is compared to the overall threshold (here this is 0.5) to see if a 0 or 1 should be predicted.

```
X = [(1,0,0),(1,1,0),(1,1,1),(1,1,1),(1,0,1),(1,0,1)]
y = [1,1,0,0,1,1]

p = perceptron(X,y)
p.initialize()
p.train()
print p.predict((1,1,1))
print p.predict((1,0,1))
```

Our X (predictors) data matrix.

Our y (target) data vector.

We instantiate our perceptron class with the data from matrix X and vector y.

The weights for the predictors are initialized (as explained previously).

We check what the perceptron would now predict given different values for the predictor variables. In the first case it will predict 0; in the second it predicts a 1.

The perceptron model is trained. It will try to train until it either converges (no more errors) or it runs out of training runs (epochs).

We'll zoom in on parts of the code that might not be so evident to grasp without further explanation. We'll start by explaining how the `train_observation()` function works. This function has two large parts. The first is to calculate the prediction of an observation and compare it to the actual value. The second part is to change the weights if the prediction seems to be wrong.

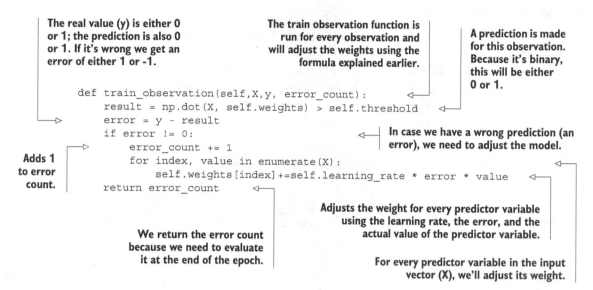

The prediction (y) is calculated by multiplying the input vector of independent variables with their respective weights and summing up the terms (as in linear regression). Then this value is compared with the threshold. If it's larger than the threshold, the algorithm will give a 1 as output, and if it's less than the threshold, the algorithm gives 0 as output. Setting the threshold is a subjective thing and depends on your business case. Let's say you're predicting whether someone has a certain lethal disease, with 1 being positive and 0 negative. In this case it's better to have a lower threshold: it's not as bad to be found positive and do a second investigation than it is to overlook the disease and let the patient die. The error is calculated, which will give the direction to the change of the weights.

```
result = np.dot(X, self.weights) > self.threshold
error = y - result
```

The weights are changed according to the sign of the error. The update is done with the learning rule for perceptrons. For every weight in the weight vector, you update its value with the following rule:

$$\Delta w_i = \alpha \varepsilon x_i$$

where Δw_i is the amount that the weight needs to be changed, α is the learning rate, ε is the error, and x_i is the ith value in the input vector (the ith predictor variable). The error count is a variable to keep track of how many observations are wrongly predicted in this epoch and is returned to the calling function. You add one observation to the error counter if the original prediction was wrong. An *epoch* is a single training run through all the observations.

```
if error != 0:
        error_count += 1
            for index, value in enumerate(X):
                self.weights[index] +=  self.learning_rate * error * value
```

The second function that we'll discuss in more detail is the train() function. This function has an internal loop that keeps on training the perceptron until it can either predict perfectly or until it has reached a certain number of training rounds (epochs), as shown in the following listing.

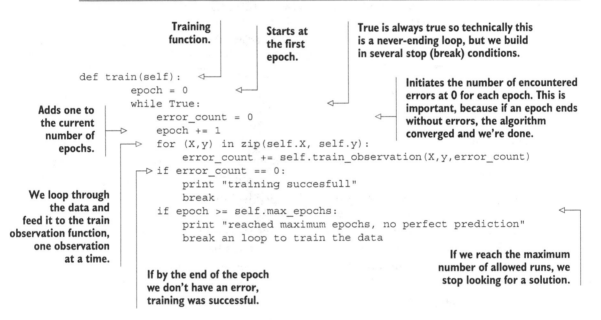

Listing 4.2   Using train functions

Most online algorithms can also handle mini-batches; this way, you can feed them batches of 10 to 1,000 observations at once while using a sliding window to go over your data. You have three options:

- *Full batch learning (also called statistical learning)*—Feed the algorithm all the data at once. This is what we did in chapter 3.
- *Mini-batch learning*—Feed the algorithm a spoonful (100, 1000, ..., depending on what your hardware can handle) of observations at a time.
- *Online learning*—Feed the algorithm one observation at a time.

Online learning techniques are related to *streaming algorithms*, where you see every data point only once. Think about incoming Twitter data: it gets loaded into the algorithms, and then the observation (tweet) is discarded because the sheer number of incoming tweets of data might soon overwhelm the hardware. Online learning algorithms differ from streaming algorithms in that they can see the same observations multiple times. True, the online learning algorithms and streaming algorithms can *both* learn from observations one by one. Where they differ is that *online algorithms* are also used on a static data source as well as on a streaming data source by presenting the data in small batches (as small as a single observation), which enables you to go over the data multiple times. This isn't the case with a *streaming algorithm*, where data flows into the system and you need to do the calculations typically immediately. They're similar in that they handle only a few at a time.

DIVIDING A LARGE MATRIX INTO MANY SMALL ONES

Whereas in the previous chapter we barely needed to deal with how exactly the algorithm estimates parameters, diving into this might sometimes help. By cutting a large data table into small matrices, for instance, we can still do a linear regression. The logic behind this matrix splitting and how a linear regression can be calculated with matrices can be found in the sidebar. It suffices to know for now that the Python libraries we're about to use will take care of the matrix splitting, and linear regression variable weights can be calculated using matrix calculus.

Block matrices and matrix formula of linear regression coefficient estimation

Certain algorithms can be translated into algorithms that use blocks of matrices instead of full matrices. When you partition a matrix into a block matrix, you divide the full matrix into parts and work with the smaller parts instead of the full matrix. In this case you can load smaller matrices into memory and perform calculations, thereby avoiding an out-of-memory error. Figure 4.4 shows how you can rewrite matrix addition A + B into submatrices.

$$
A + B =
\begin{bmatrix} a_{1,1} & \cdots & a_{1,m} \\ \vdots & \ddots & \vdots \\ a_{n,1} & \cdots & a_{n,m} \end{bmatrix}
+
\begin{bmatrix} b_{1,1} & \cdots & b_{1,m} \\ \vdots & \ddots & \vdots \\ b_{n,1} & \cdots & b_{n,m} \end{bmatrix}
$$

$$
=
\begin{bmatrix} a_{1,1} & \cdots & a_{1,m} \\ \vdots & \ddots & \vdots \\ a_{j,1} & \cdots & a_{j,m} \\ a_{j+1,1} & \cdots & a_{j+1,m} \\ \vdots & \ddots & \vdots \\ a_{n,1} & \cdots & a_{n,m} \end{bmatrix}
+
\begin{bmatrix} b_{1,1} & \cdots & b_{1,m} \\ \vdots & \ddots & \vdots \\ b_{j,1} & \cdots & b_{j,m} \\ b_{j+1,1} & \cdots & b_{j+1,m} \\ \vdots & \ddots & \vdots \\ b_{n,1} & \cdots & b_{n,m} \end{bmatrix}
=
\begin{bmatrix} A_1 \\ A_2 \end{bmatrix}
+
\begin{bmatrix} B_1 \\ B_2 \end{bmatrix}
$$

Figure 4.4 Block matrices can be used to calculate the sum of the matrices A and B.

(continued)

The formula in figure 4.4 shows that there's no difference between adding matrices A and B together in one step or first adding the upper half of the matrices and then adding the lower half.

All the common matrix and vector operations, such as multiplication, inversion, and singular value decomposition (a variable reduction technique like PCA), can be written in terms of block matrices.[1] Block matrix operations save memory by splitting the problem into smaller blocks and are easy to parallelize.

Although most numerical packages have highly optimized code, they work only with matrices that can fit into memory and will use block matrices in memory when advantageous. With out-of-memory matrices, they don't optimize this for you and it's up to you to partition the matrix into smaller matrices and to implement the block matrix version.

A *linear regression* is a way to predict continuous variables with a linear combination of its predictors; one of the most basic ways to perform the calculations is with a technique called *ordinary least squares*. The formula in matrix form is

$$\beta = (X^TX)^{-1}X^Ty$$

where β is the coefficients you want to retrieve, X is the predictors, and y is the target variable.

The Python tools we have at our disposal to accomplish our task are the following:

- *bcolz* is a Python library that can store data arrays compactly and uses the hard drive when the array no longer fits into the main memory.
- *Dask* is a library that enables you to optimize the flow of calculations and makes performing calculations in parallel easier. It doesn't come packaged with the default Anaconda setup so make sure to use `conda install dask` on your virtual environment before running the code below. Note: some errors have been reported on importing Dask when using 64bit Python. Dask is dependent on a few other libraries (such as toolz), but the dependencies should be taken care of automatically by pip or conda.

The following listing demonstrates block matrix calculations with these libraries.

---

[1] For those who want to give it a try, Given transformations are easier to achieve than Householder transformations when calculating singular value decompositions.

Listing 4.3   Block matrix calculations with bcolz and Dask libraries

Number of observations
(scientific notation).
1e4 = 10.000. Feel
free to change this.

Creates fake data: np.arange(n).reshape(n/2,2) creates
a matrix of 5000 by 2 (because we set n to 10.000).
bc.carray = numpy is an array extension that can
swap to disc. This is also stored in a compressed way.
rootdir = 'ar.bcolz' --> creates a file on disc in case out of
RAM. You can check this on your file system next to this
ipython file or whatever location you ran this code from.
mode = 'w' --> is the write mode. dtype = 'float64' --> is
the storage type of the data (which is float numbers).

```
import dask.array as da
import bcolz as bc
import numpy as np
import dask

n = 1e4

ar = bc.carray(np.arange(n).reshape(n/2,2)  , dtype='float64',
    rootdir = 'ar.bcolz', mode = 'w')
y  = bc.carray(np.arange(n/2), dtype='float64', rootdir =
    'yy.bcolz', mode = 'w')
```

```
dax = da.from_array(ar, chunks=(5,5))
dy = da.from_array(y,chunks=(5,5))
```

Block matrices are created for the predictor variables
(ar) and target (y). A block matrix is a matrix cut in
pieces (blocks). da.from_array() reads data
from disc or RAM (wherever it resides currently).
chunks=(5,5): every block is a 5x5 matrix
(unless < 5 observations or variables are left).

The XTX is defined (defining it as "lazy") as the
X matrix multiplied with its transposed version.
This is a building block of the formula to do
linear regression using matrix calculation.

Xy is the y vector multiplied with the transposed
X matrix. Again the matrix is only defined, not
calculated yet. This is also a building block of the
formula to do linear regression using matrix
calculation (see formula).

```
XTX = dax.T.dot(dax)
Xy  = dax.T.dot(dy)
```

```
coefficients = np.linalg.inv(XTX.compute()).dot(Xy.compute())
```

```
coef = da.from_array(coefficients,chunks=(5,5))
```

The coefficients are also put
into a block matrix. We got a
numpy array back from the last
step so we need to explicitly
convert it back to a "da array."

```
ar.flush()      Flush memory data. It's no longer needed
y.flush()       to have large matrices in memory.
```

```
predictions = dax.dot(coef).compute()
print predictions
```

Score the model
(make predictions).

The coefficients are calculated using the matrix
linear regression function. np.linalg.inv() is the
^(-1) in this function, or "inversion" of the
matrix. X.dot(y) --> multiplies the matrix X
with another matrix y.

Note that you don't need to use a block matrix inversion because XTX is a square matrix with size nr. of predictors * nr. of predictors. This is fortunate because Dask doesn't yet support block matrix inversion. You can find more general information on matrix arithmetic on the Wikipedia page at https://en.wikipedia.org/wiki/Matrix_ (mathematics).

MAPREDUCE

MapReduce algorithms are easy to understand with an analogy: Imagine that you were asked to count all the votes for the national elections. Your country has 25 parties, 1,500 voting offices, and 2 million people. You could choose to gather all the voting tickets from every office individually and count them centrally, or you could ask the local offices to count the votes for the 25 parties and hand over the results to you, and you could then aggregate them by party.

Map reducers follow a similar process to the second way of working. They first map values to a key and then do an aggregation on that key during the reduce phase. Have a look at the following listing's pseudo code to get a better feeling for this.

> **Listing 4.4 MapReduce pseudo code example**

```
For each person in voting office:
    Yield (voted_party, 1)
For each vote in voting office:
    add_vote_to_party()
```

One of the advantages of MapReduce algorithms is that they're easy to parallelize and distribute. This explains their success in distributed environments such as Hadoop, but they can also be used on individual computers. We'll take a more in-depth look at them in the next chapter, and an example (JavaScript) implementation is also provided in chapter 9. When implementing MapReduce in Python, you don't need to start from scratch. A number of libraries have done most of the work for you, such as Hadoopy, Octopy, Disco, or Dumbo.

4.2.2 *Choosing the right data structure*

Algorithms can make or break your program, but the way you store your data is of equal importance. Data structures have different storage requirements, but also influence the performance of *CRUD* (create, read, update, and delete) and other operations on the data set.

Figure 4.5 shows you have many different data structures to choose from, three of which we'll discuss here: sparse data, tree data, and hash data. Let's first have a look at sparse data sets.

SPARSE DATA

A sparse data set contains relatively little information compared to its entries (observations). Look at figure 4.6: almost everything is "0" with just a single "1" present in the second observation on variable 9.

Data like this might look ridiculous, but this is often what you get when converting textual data to binary data. Imagine a set of 100,000 completely unrelated Twitter

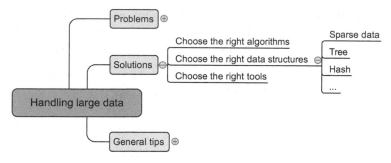

Figure 4.5 Overview of data structures often applied in data science when working with large data

	1	2	3	4	5	6	7	8	9	10	11	12	13	14	15	16
1	0	0	0	0	0	0	0	0	0	0	0	0	0	0	0	0
2	0	0	0	0	0	0	0	0	1	0	0	0	0	0	0	0
3	0	0	0	0	0	0	0	0	0	0	0	0	0	0	0	0
4	0	0	0	0	0	0	0	0	0	0	0	0	0	0	0	0

Figure 4.6 Example of a sparse matrix: almost everything is 0; other values are the exception in a sparse matrix

tweets. Most of them probably have fewer than 30 words, but together they might have hundreds or thousands of distinct words. In the chapter on text mining we'll go through the process of cutting text documents into words and storing them as vectors. But for now imagine what you'd get if every word was converted to a binary variable, with "1" representing "present in this tweet," and "0" meaning "not present in this tweet." This would result in sparse data indeed. The resulting large matrix can cause memory problems even though it contains little information.

Luckily, data like this can be stored compacted. In the case of figure 4.6 it could look like this:

```
data = [(2,9,1)]
```

Row 2, column 9 holds the value 1.

Support for working with sparse matrices is growing in Python. Many algorithms now support or return sparse matrices.

TREE STRUCTURES

Trees are a class of data structure that allows you to retrieve information much faster than scanning through a table. A tree always has a root value and subtrees of children, each with its children, and so on. Simple examples would be your own family tree or a biological tree and the way it splits into branches, twigs, and leaves. Simple decision rules make it easy to find the child tree in which your data resides. Look at figure 4.7 to see how a tree structure enables you to get to the relevant information quickly.

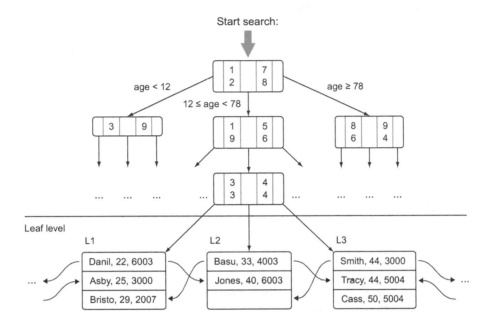

Figure 4.7 Example of a tree data structure: decision rules such as age categories can be used to quickly locate a person in a family tree

In figure 4.7 you start your search at the top and first choose an age category, because apparently that's the factor that cuts away the most alternatives. This goes on and on until you get what you're looking for. For whoever isn't acquainted with the Akinator, we recommend visiting http://en.akinator.com/. The Akinator is a djinn in a magical lamp that tries to guess a person in your mind by asking you a few questions about him or her. Try it out and be amazed . . . or see how this magic is a tree search.

Trees are also popular in databases. Databases prefer not to scan the table from the first line until the last, but to use a device called an *index* to avoid this. Indices are often based on data structures such as trees and hash tables to find observations faster. The use of an index speeds up the process of finding data enormously. Let's look at these hash tables.

HASH TABLES

Hash tables are data structures that calculate a key for every value in your data and put the keys in a bucket. This way you can quickly retrieve the information by looking in the right bucket when you encounter the data. Dictionaries in Python are a hash table implementation, and they're a close relative of key-value stores. You'll encounter them in the last example of this chapter when you build a recommender system within a database. Hash tables are used extensively in databases as indices for fast information retrieval.

4.2.3 Selecting the right tools

With the right class of algorithms and data structures in place, it's time to choose the right tool for the job. The right tool can be a Python library or at least a tool that's controlled from Python, as shown figure 4.8. The number of helpful tools available is enormous, so we'll look at only a handful of them.

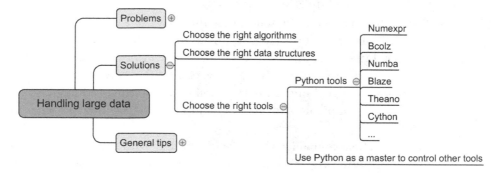

Figure 4.8   Overview of tools that can be used when working with large data

PYTHON TOOLS

Python has a number of libraries that can help you deal with large data. They range from smarter data structures over code optimizers to just-in-time compilers. The following is a list of libraries we like to use when confronted with large data:

- *Cython*—The closer you get to the actual hardware of a computer, the more vital it is for the computer to know what types of data it has to process. For a computer, adding 1 + 1 is different from adding 1.00 + 1.00. The first example consists of integers and the second consists of floats, and these calculations are performed by different parts of the CPU. In Python you don't have to specify what data types you're using, so the Python compiler has to infer them. But inferring data types is a slow operation and is partially why Python isn't one of the fastest languages available. Cython, a superset of Python, solves this problem by forcing the programmer to specify the data type while developing the program. Once the compiler has this information, it runs programs much faster. See http://cython.org/ for more information on Cython.
- *Numexpr*—Numexpr is at the core of many of the big data packages, as is NumPy for in-memory packages. Numexpr is a numerical expression evaluator for NumPy but can be many times faster than the original NumPy. To achieve

this, it rewrites your expression and uses an internal (just-in-time) compiler. See https://github.com/pydata/numexpr for details on Numexpr.

- *Numba*—Numba helps you to achieve greater speed by compiling your code right before you execute it, also known as *just-in-time compiling*. This gives you the advantage of writing high-level code but achieving speeds similar to those of C code. Using Numba is straightforward; see http://numba.pydata.org/.

- *Bcolz*—Bcolz helps you overcome the out-of-memory problem that can occur when using NumPy. It can store and work with arrays in an optimal compressed form. It not only slims down your data need but also uses Numexpr in the background to reduce the calculations needed when performing calculations with bcolz arrays. See http://bcolz.blosc.org/.

- *Blaze*—Blaze is ideal if you want to use the power of a database backend but like the "Pythonic way" of working with data. Blaze will translate your Python code into SQL but can handle many more data stores than relational databases such as CSV, Spark, and others. Blaze delivers a unified way of working with many databases and data libraries. Blaze is still in development, though, so many features aren't implemented yet. See http://blaze.readthedocs.org/en/latest/index.html.

- *Theano*—Theano enables you to work directly with the graphical processing unit (GPU) and do symbolical simplifications whenever possible, and it comes with an excellent just-in-time compiler. On top of that it's a great library for dealing with an advanced but useful mathematical concept: tensors. See http://deeplearning.net/software/theano/.

- *Dask*—Dask enables you to optimize your flow of calculations and execute them efficiently. It also enables you to distribute calculations. See http://dask.pydata.org/en/latest/.

These libraries are mostly about using Python itself for data processing (apart from Blaze, which also connects to databases). To achieve high-end performance, you can use Python to communicate with all sorts of databases or other software.

USE PYTHON AS A MASTER TO CONTROL OTHER TOOLS

Most software and tool producers support a Python interface to their software. This enables you to tap into specialized pieces of software with the ease and productivity that comes with Python. This way Python sets itself apart from other popular data science languages such as R and SAS. You should take advantage of this luxury and exploit the power of specialized tools to the fullest extent possible. Chapter 6 features a case study using Python to connect to a NoSQL database, as does chapter 7 with graph data.

Let's now have a look at more general helpful tips when dealing with large data.

4.3   General programming tips for dealing with large data sets

The tricks that work in a general programming context still apply for data science. Several might be worded slightly differently, but the principles are essentially the same for all programmers. This section recapitulates those tricks that are important in a data science context.

You can divide the general tricks into three parts, as shown in the figure 4.9 mind map:

- *Don't reinvent the wheel.* Use tools and libraries developed by others.
- *Get the most out of your hardware.* Your machine is never used to its full potential; with simple adaptions you can make it work harder.
- *Reduce the computing need.* Slim down your memory and processing needs as much as possible.

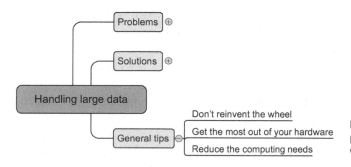

Figure 4.9   Overview of general programming best practices when working with large data

"Don't reinvent the wheel" is easier said than done when confronted with a specific problem, but your first thought should always be, 'Somebody else must have encountered this same problem before me.'

4.3.1   Don't reinvent the wheel

"Don't repeat anyone" is probably even better than "don't repeat yourself." Add value with your actions: make sure that they matter. Solving a problem that has already been solved is a waste of time. As a data scientist, you have two large rules that can help you deal with large data and make you much more productive, to boot:

- *Exploit the power of databases.* The first reaction most data scientists have when working with large data sets is to prepare their analytical base tables inside a database. This method works well when the features you want to prepare are fairly simple. When this preparation involves advanced modeling, find out if it's possible to employ user-defined functions and procedures. The last example of this chapter is on integrating a database into your workflow.
- *Use optimized libraries.* Creating libraries like Mahout, Weka, and other machine-learning algorithms requires time and knowledge. They are highly optimized

and incorporate best practices and state-of-the art technologies. Spend your time on getting things done, not on reinventing and repeating others people's efforts, unless it's for the sake of understanding how things work.

Then you must consider your hardware limitation.

4.3.2   Get the most out of your hardware

Resources on a computer can be idle, whereas other resources are over-utilized. This slows down programs and can even make them fail. Sometimes it's possible (and necessary) to shift the workload from an overtaxed resource to an underutilized resource using the following techniques:

- *Feed the CPU compressed data.* A simple trick to avoid CPU starvation is to feed the CPU compressed data instead of the inflated (raw) data. This will shift more work from the hard disk to the CPU, which is exactly what you want to do, because a hard disk can't follow the CPU in most modern computer architectures.
- *Make use of the GPU.* Sometimes your CPU and not your memory is the bottleneck. If your computations are parallelizable, you can benefit from switching to the GPU. This has a much higher throughput for computations than a CPU. The GPU is enormously efficient in parallelizable jobs but has less cache than the CPU. But it's pointless to switch to the GPU when your hard disk is the problem. Several Python packages, such as Theano and NumbaPro, will use the GPU without much programming effort. If this doesn't suffice, you can use a CUDA (Compute Unified Device Architecture) package such as PyCUDA. It's also a well-known trick in bitcoin mining, if you're interested in creating your own money.
- *Use multiple threads.* It's still possible to parallelize computations on your CPU. You can achieve this with normal Python threads.

4.3.3   Reduce your computing needs

"Working smart + hard = achievement." This also applies to the programs you write. The best way to avoid having large data problems is by removing as much of the work as possible up front and letting the computer work only on the part that can't be skipped. The following list contains methods to help you achieve this:

- *Profile your code and remediate slow pieces of code.* Not every piece of your code needs to be optimized; use a profiler to detect slow parts inside your program and remediate these parts.
- *Use compiled code whenever possible, certainly when loops are involved.* Whenever possible use functions from packages that are optimized for numerical computations instead of implementing everything yourself. The code in these packages is often highly optimized and compiled.
- *Otherwise, compile the code yourself.* If you can't use an existing package, use either a just-in-time compiler or implement the slowest parts of your code in a

lower-level language such as C or Fortran and integrate this with your codebase. If you make the step to *lower-level languages* (languages that are closer to the universal computer bytecode), learn to work with computational libraries such as LAPACK, BLAST, Intel MKL, and ATLAS. These are highly optimized, and it's difficult to achieve similar performance to them.

- *Avoid pulling data into memory.* When you work with data that doesn't fit in your memory, avoid pulling everything into memory. A simple way of doing this is by reading data in chunks and parsing the data on the fly. This won't work on every algorithm but enables calculations on extremely large data sets.

- *Use generators to avoid intermediate data storage.* Generators help you return data per observation instead of in batches. This way you avoid storing intermediate results.

- *Use as little data as possible.* If no large-scale algorithm is available and you aren't willing to implement such a technique yourself, then you can still train your data on only a sample of the original data.

- *Use your math skills to simplify calculations as much as possible.* Take the following equation, for example: $(a + b)^2 = a^2 + 2ab + b^2$. The left side will be computed much faster than the right side of the equation; even for this trivial example, it could make a difference when talking about big chunks of data.

4.4 Case study 1: Predicting malicious URLs

The internet is probably one of the greatest inventions of modern times. It has boosted humanity's development, but not everyone uses this great invention with honorable intentions. Many companies (Google, for one) try to protect us from fraud by detecting malicious websites for us. Doing so is no easy task, because the internet has billions of web pages to scan. In this case study we'll show how to work with a data set that no longer fits in memory.

What we'll use

- *Data*—The data in this case study was made available as part of a research project. The project contains data from 120 days, and each observation has approximately 3,200,000 features. The target variable contains 1 if it's a malicious website and -1 otherwise. For more information, please see "Beyond Blacklists: Learning to Detect Malicious Web Sites from Suspicious URLs."[2]

- *The Scikit-learn library*—You should have this library installed in your Python environment at this point, because we used it in the previous chapter.

As you can see, we won't be needing much for this case, so let's dive into it.

---

[2] Justin Ma, Lawrence K. Saul, Stefan Savage, and Geoffrey M. Voelker, "Beyond Blacklists: Learning to Detect Malicious Web Sites from Suspicious URLs," Proceedings of the ACM SIGKDD Conference, Paris (June 2009), 1245–53.

4.4.1   Step 1: Defining the research goal

The goal of our project is to detect whether certain URLs can be trusted or not. Because the data is so large we aim to do this in a memory-friendly way. In the next step we'll first look at what happens if we don't concern ourselves with memory (RAM) issues.

4.4.2   Step 2: Acquiring the URL data

Start by downloading the data from http://sysnet.ucsd.edu/projects/url/#datasets and place it in a folder. Choose the data in SVMLight format. SVMLight is a text-based format with one observation per row. To save space, it leaves out the zeros.

```
---------------------------------------------------------------------
MemoryError                             Traceback (most recent call last)
<ipython-input-532-d196c05088ce> in <module>()
      5 print "there are %d files" % len(files)
      6 X,y = load_svmlight_file(files[0] ,n_features=3500000)
----> 7 X.todense()
```

Figure 4.10   Memory error when trying to take a large data set into memory

The following listing and figure 4.10 show what happens when you try to read in 1 file out of the 120 and create the normal matrix as most algorithms expect. The todense() method changes the data from a special file format to a normal matrix where every entry contains a value.

> **Listing 4.5   Generating an out-of-memory error**

```
import glob
from sklearn.datasets import load_svmlight_file
files = glob.glob('C:\Users\Gebruiker\Downloads\       ← Points to files (Linux).
url_svmlight.tar\url_svmlight\*.svm')
files = glob.glob('C:\Users\Gebruiker\Downloads\       ← Points to files (Windows:
url_svmlight\url_svmlight\*.svm')                         tar file needs to be
print "there are %d files" % len(files)                  untarred first).
X,y = load_svmlight_file(files[0],n_features=3231952)  ← Indication of number of files.
X.todense()                                            ← Loads files.
```

The data is a big, but sparse, matrix. By turning it into a dense matrix (every 0 is represented in the file), we create an out-of-memory error.

Surprise, surprise, we get an out-of-memory error. That is, unless you run this code on a huge machine. After a few tricks you'll no longer run into these memory problems and will detect 97% of the malicious sites.

TOOLS AND TECHNIQUES

We ran into a memory error while loading a single file—still 119 to go. Luckily, we have a few tricks up our sleeve. Let's try these techniques over the course of the case study:

- Use a sparse representation of data.
- Feed the algorithm compressed data instead of raw data.
- Use an online algorithm to make predictions.

We'll go deeper into each "trick" when we get to use it. Now that we have our data locally, let's access it. Step 3 of the data science process, data preparation and cleansing, isn't necessary in this case because the URLs come pre-cleaned. We'll need a form of exploration before unleashing our learning algorithm, though.

4.4.3 *Step 4: Data exploration*

To see if we can even apply our first trick (sparse representation), we need to find out whether the data does indeed contain lots of zeros. We can check this with the following piece of code:

```
print "number of non-zero entries %2.6f"  % float((X.nnz)/(float(X.shape[0])
* float(X.shape[1])))
```

This outputs the following:

```
number of non-zero entries 0.000033
```

Data that contains little information compared to zeros is called *sparse data*. This can be saved more compactly if you store the data as `[(0,0,1),(4,4,1)]` instead of

```
[[1,0,0,0,0],[0,0,0,0,0],[0,0,0,0,0],[0,0,0,0,0],[0,0,0,0,1]]
```

One of the file formats that implements this is SVMLight, and that's exactly why we downloaded the data in this format. We're not finished yet, though, because we need to get a feel of the dimensions within the data.

   To get this information we already need to keep the data compressed while checking for the maximum number of observations and variables. We also need to *read in data file by file*. This way you consume even less memory. A second trick is to feed the CPU compressed files. In our example, it's already packed in the tar.gz format. You unpack a file only when you need it, without writing it to the hard disk (the slowest part of your computer).

   For our example, shown in listing 4.6, we'll only work on the first 5 files, but feel free to use all of them.

Listing 4.6   Checking data size

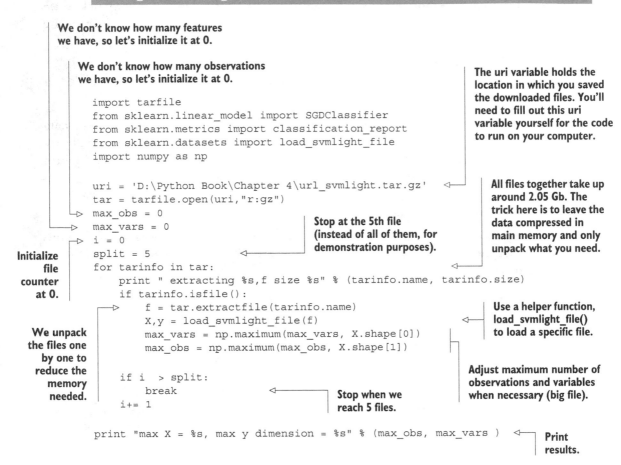

We don't know how many features
we have, so let's initialize it at 0.

We don't know how many observations
we have, so let's initialize it at 0.

The uri variable holds the location in which you saved the downloaded files. You'll need to fill out this uri variable yourself for the code to run on your computer.

```
import tarfile
from sklearn.linear_model import SGDClassifier
from sklearn.metrics import classification_report
from sklearn.datasets import load_svmlight_file
import numpy as np

uri = 'D:\Python Book\Chapter 4\url_svmlight.tar.gz'
tar = tarfile.open(uri,"r:gz")
max_obs = 0
max_vars = 0
i = 0
split = 5
for tarinfo in tar:
    print " extracting %s,f size %s" % (tarinfo.name, tarinfo.size)
    if tarinfo.isfile():
        f = tar.extractfile(tarinfo.name)
        X,y = load_svmlight_file(f)
        max_vars = np.maximum(max_vars, X.shape[0])
        max_obs = np.maximum(max_obs, X.shape[1])

    if i  > split:
        break
    i+= 1

print "max X = %s, max y dimension = %s" % (max_obs, max_vars )
```

Initialize file counter at 0.

Stop at the 5th file (instead of all of them, for demonstration purposes).

All files together take up around 2.05 Gb. The trick here is to leave the data compressed in main memory and only unpack what you need.

We unpack the files one by one to reduce the memory needed.

Use a helper function, load_svmlight_file() to load a specific file.

Adjust maximum number of observations and variables when necessary (big file).

Stop when we reach 5 files.

Print results.

Part of the code needs some extra explanation. In this code we loop through the svm files inside the tar archive. We unpack the files one by one to reduce the memory needed. As these files are in the SVM format, we use a helper, function load_svmlight _file() to load a specific file. Then we can see how many observations and variables the file has by checking the shape of the resulting data set.

Armed with this information we can move on to model building.

4.4.4   *Step 5: Model building*

Now that we're aware of the dimensions of our data, we can apply the same two tricks (sparse representation of compressed file) and add the third (using an online algorithm), in the following listing. Let's find those harmful websites!

Listing 4.7   Creating a model to distinguish the malicious from the normal URLs

We know number of features from data exploration.

The target variable can be 1 or -1. "1": website safe to visit, "-1": website unsafe.

Set up stochastic gradient classifier.

```
classes = [-1,1]
sgd = SGDClassifier(loss="log")
n_features=3231952
split = 5
i = 0
for tarinfo in tar:
    if i > split:
        break
    if tarinfo.isfile():
        f = tar.extractfile(tarinfo.name)
        X,y = load_svmlight_file(f,n_features=n_features)
        if i < split:
            sgd.partial_fit(X, y, classes=classes)
        if i == split:
            print classification_report(sgd.predict(X),y)
    i += 1
```

All files together take up around 2.05 Gb. The trick here is to leave data compressed in main memory and only unpack what you need.

We unpack the files one by one to reduce the memory needed.

Use a helper function, load_svmlight_file() to load a specific file.

Initialize file counter at 0.

Stop when we reach 5 files and print results.

Third important thing is online algorithm. It can be fed data points file by file (batches).

Stop at 5th file (instead of all of them, for demonstration purposes).

The code in the previous listing looks fairly similar to what we did before, apart from the stochastic gradient descent classifier SGDClassifier().

Here, we trained the algorithm iteratively by presenting the observations in one file with the partial_fit() function.

Looping through only the first 5 files here gives the output shown in table 4.1. The table shows classification diagnostic measures: precision, recall, F1-score, and support.

Table 4.1   Classification problem: Can a website be trusted or not?

	precision	recall	f1-score	support
-1	0.97	0.99	0.98	14045
1	0.97	0.94	0.96	5955
avg/total	0.97	0.97	0.97	20000

Only 3% (1 - 0.97) of the malicious sites aren't detected (*precision*), and 6% (1 - 0.94) of the sites detected are falsely accused (*recall*). This is a decent result, so we can conclude that the methodology works. If we rerun the analysis, the result might be slightly different, because the algorithm could converge slightly differently. If you don't mind waiting a while, you can go for the full data set. You can now handle all the data without problems. We won't have a sixth step (presentation or automation) in this case study.

Now let's look at a second application of our techniques; this time you'll build a recommender system inside a database. For a well-known example of recommender systems visit the Amazon website. While browsing, you'll soon be confronted with recommendations: "People who bought this product also bought..."

4.5   *Case study 2: Building a recommender system inside a database*

In reality most of the data you work with is stored in a relational database, but most databases aren't suitable for data mining. But as shown in this example, it's possible to adapt our techniques so you can do a large part of the analysis inside the database itself, thereby profiting from the database's query optimizer, which will optimize the code for you. In this example we'll go into how to use the hash table data structure and how to use Python to control other tools.

4.5.1   *Tools and techniques needed*

Before going into the case study we need to have a quick look at the required tools and theoretical background to what we're about to do here.

TOOLS

- *MySQL database*—Needs a MySQL database to work with. If you haven't installed a MySQL community server, you can download one from www.mysql.com. Appendix C: "Installing a MySQL server" explains how to set it up.
- *MySQL database connection Python library*—To connect to this server from Python you'll also need to install SQLAlchemy or another library capable of communicating with MySQL. We're using MySQLdb. On Windows you can't use Conda right off the bat to install it. First install Binstar (another package management service) and look for the appropriate mysql-python package for your Python setup.

```
conda install binstar
binstar search -t conda mysql-python
```

The following command entered into the Windows command line worked for us (after activating the Python environment):

```
conda install --channel https://conda.binstar.org/krisvanneste mysql-python
```

Again, feel free to go for the SQLAlchemy library if that's something you're more comfortable with.

- We will also need the *pandas* python library, but that should already be installed by now.

With the infrastructure in place, let's dive into a few of the techniques.

TECHNIQUES

A simple recommender system will look for customers who've rented similar movies as you have and then suggest those that the others have watched but you haven't seen yet. This technique is called *k-nearest neighbors* in machine learning.

A customer who behaves similarly to you isn't necessarily *the* most similar customer. You'll use a technique to ensure that you can find similar customers (local optima) without guarantees that you've found the best customer (global optimum). A common technique used to solve this is called *Locality-Sensitive Hashing*. A good overview of papers on this topic can be found at http://www.mit.edu/~andoni/LSH/.

The idea behind Locality-Sensitive Hashing is simple: Construct functions that map similar customers close together (they're put in a bucket with the same label) and make sure that objects that are different are put in different buckets.

Central to this idea is a function that performs the mapping. This function is called a hash function: a function that maps any range of input to a fixed output. The simplest hash function concatenates the values from several random columns. It doesn't matter how many columns (scalable input); it brings it back to a single column (fixed output).

You'll set up three hash functions to find similar customers. The three functions take the values of three movies:

- The first function takes the values of movies 10, 15, and 28.
- The second function takes the values of movies 7, 18, and 22.
- The last function takes the values of movies 16, 19, and 30.

This will ensure that the customers who are in the same bucket share at least several movies. But the customers inside one bucket might still differ on the movies that weren't included in the hashing functions. To solve this you still need to compare the customers within the bucket with each other. For this you need to create a new distance measure.

The distance that you'll use to compare customers is called the hamming distance. The hamming distance is used to calculate how much two strings differ. The distance is defined as the number of different characters in a string. Table 4.2 offers a few examples of the hamming distance.

Table 4.2   Examples of calculating the hamming distance

String 1	String 2	Hamming distance
Hat	Cat	1
Hat	Mad	2
Tiger	Tigre	2
Paris	Rome	5

Comparing multiple columns is an expensive operation, so you'll need a trick to speed this up. Because the columns contain a binary (0 or 1) variable to indicate whether a customer has bought a movie or not, you can concatenate the information so that the same information is contained in a new column. Table 4.3 shows the "movies" variable that contains as much information as all the movie columns combined.

Table 4.3 Combining the information from different columns into the movies column. This is also how DNA works: all information in a long string.

Column 1	Movie 1	Movie 2	Movie 3	Movie 4	movies
Customer 1	1	0	1	1	1011
Customer 2	0	0	0	1	0001

This allows you to calculate the hamming distance much more efficiently. By handling this operator as a bit, you can exploit the XOR operator. The outcome of the XOR operator ($^\wedge$) is as follows:

```
1^1 = 0
1^0 = 1
0^1 = 1
0^0 = 0
```

With this in place, the process to find similar customers becomes very simple. Let's first look at it in pseudo code:

Preprocessing:

1 Define p (for instance, 3) functions that select k (for instance, 3) entries from the vector of movies. Here we take 3 functions (p) that each take 3 (k) movies.
2 Apply these functions to every point and store them in a separate column. (In literature each function is called a hash function and each column will store a bucket.)

Querying point q:

1 Apply the same p functions to the point (observation) q you want to query.
2 Retrieve for every function the points that correspond to the result in the corresponding bucket.
   Stop when you've retrieved all the points in the buckets or reached 2p points (for example 10 if you have 5 functions).
3 Calculate the distance for each point and return the points with the minimum distance.

Let's look at an actual implementation in Python to make this all clearer.

4.5.2   Step 1: Research question

Let's say you're working in a video store and the manager asks you if it's possible to use the information on what movies people rent to predict what other movies they might like. Your boss has stored the data in a MySQL database, and it's up to you to do the analysis. What he is referring to is a recommender system, an automated system that learns people's preferences and recommends movies and other products the customers haven't tried yet. The goal of our case study is to create a memory-friendly recommender system. We'll achieve this using a database and a few extra tricks. We're going to create the data ourselves for this case study so we can skip the data retrieval step and move right into data preparation. And after that we can skip the data exploration step and move straight into model building.

4.5.3   Step 3: Data preparation

The data your boss has collected is shown in table 4.4. We'll create this data ourselves for the sake of demonstration.

Table 4.4   Excerpt from the client database and the movies customers rented

Customer	Movie 1	Movie 2	Movie 3	...	Movie 32
Jack Dani	1	0	0		1
Wilhelmson	1	1	0		1
...					
Jane Dane	0	0	1		0
Xi Liu	0	0	0		1
Eros Mazo	1	1	0		1
...					

For each customer you get an indication of whether they've rented the movie before (1) or not (0). Let's see what else you'll need so you can give your boss the recommender system he desires.

First let's connect Python to MySQL to create our data. Make a connection to MySQL using your username and password. In the following listing we used a database called "test". Replace the user, password, and database name with the appropriate values for your setup and retrieve the connection and the cursor. A database cursor is a control structure that remembers where you are currently in the database.

Listing 4.8 Creating customers in the database

```
import MySQLdb
import pandas as pd

user = '****'
password = '****'
database = 'test'
mc = MySQLdb.connect('localhost',user,password,database)
cursor = mc.cursor()

nr_customers = 100
colnames = ["movie%d" %i for i in range(1,33)]
pd.np.random.seed(2015)
generated_customers = pd.np.random.randint(0,2,32 *
    nr_customers).reshape(nr_customers,32)

data = pd.DataFrame(generated_customers, columns = list(colnames))
data.to_sql('cust',mc, flavor = 'mysql', index = True, if_exists =
    'replace', index_label = 'cust_id')
```

First we establish the connection; you'll need to fill out your own username, password, and schema-name (variable "database").

Next we simulate a database with customers and create a few observations.

Store the data inside a Pandas data frame and write the data frame in a MySQL table called "cust". If this table already exists, replace it.

We create 100 customers and randomly assign whether they did or didn't see a certain movie, and we have 32 movies in total. The data is first created in a Pandas data frame but is then turned into SQL code. Note: You might run across a warning when running this code. The warning states: *The "mysql" flavor with DBAPI connection is deprecated and will be removed in future versions. MySQL will be further supported with SQLAlchemy engines.* Feel free to already switch to SQLAlchemy or another library. We'll use SQLAlchemy in other chapters, but used MySQLdb here to broaden the examples.

To efficiently query our database later on we'll need additional data preparation, including the following things:

- Creating bit strings. The bit strings are compressed versions of the columns' content (0 and 1 values). First these binary values are concatenated; then the resulting bit string is reinterpreted as a number. This might sound abstract now but will become clearer in the code.
- Defining hash functions. The hash functions will in fact create the bit strings.
- Adding an index to the table, to quicken data retrieval.

CREATING BIT STRINGS

Now you make an intermediate table suited for querying, apply the hash functions, and represent the sequence of bits as a decimal number. Finally, you can place them in a table.

First, you need to create bit strings. You need to convert the string "11111111" to a binary or a numeric value to make the hamming function work. We opted for a numeric representation, as shown in the next listing.

Listing 4.9 Creating bit strings

> We represent the string as a numeric value. The string will be a concatenation of zeros (0) and ones (1) because these indicate whether someone has seen a certain movie or not. The strings are then regarded as bit code. For example: 0011 is the same as the number 3. What def createNum() does: takes in 8 values, concatenates these 8 column values and turns them into a string, then turns the byte code of the string into a number.

```
def createNum(x1,x2,x3,x4,x5,x6,x7,x8):
    return   [int('%d%d%d%d%d%d%d%d' % (i1,i2,i3,i4,i5,i6,i7,i8),2)
for (i1,i2,i3,i4,i5,i6,i7,i8) in zip(x1,x2,x3,x4,x5,x6,x7,x8)]

assert int('1111',2) == 15
assert int('1100',2) == 12
assert createNum([1,1],[1,1],[1,1],[1,1],[1,1],[1,1],[1,0],[1,0])
    == [255,252]

store = pd.DataFrame()
store['bit1'] = createNum(data.movie1,
    data.movie2,data.movie3,data.movie4,data.movie5,
data.movie6,data.movie7,data.movie8)
store['bit2'] = createNum(data.movie9,
    data.movie10,data.movie11,data.movie12,data.movie13,
data.movie14,data.movie15,data.movie16)
store['bit3'] = createNum(data.movie17,
    data.movie18,data.movie19,data.movie20,data.movie21,
data.movie22,data.movie23,data.movie24)
store['bit4'] = createNum(data.movie25,
    data.movie26,data.movie27,data.movie28,data.movie29,
data.movie30,data.movie31,data.movie32)
```

Translate the movie column to 4 bit strings in numeric form. Each bit string represents 8 movies. 4*8 = 32 movies. Note: you could use a 32-bit string instead of 4*8 to keep the code short.

Test if the function works correctly. Binary code 1111 is the same as 15 (=1*8+1*4+1*2+1*1). If the assert fails, it will raise an assert error; otherwise nothing will happen.

By converting the information of 32 columns into 4 numbers, we compressed it for later lookup. Figure 4.11 shows what we get when asking for the first 2 observations (customer movie view history) in this new format.

```
store[0:2]
```

The next step is to create the hash functions, because they'll enable us to sample the data we'll use to determine whether two customers have similar behavior.

	bit1	bit2	bit3	bit4
0	10	62	42	182
1	23	28	223	180

Figure 4.11 First 2 customers' information on all 32 movies after bit string to numeric conversion

CREATING A HASH FUNCTION

The hash functions we create take the values of movies for a customer. We decided in the theory part of this case study to create 3 hash functions: the first function combines the movies 10, 5, and 18; the second combines movies 7, 18, and 22; and the third one combines 16, 19, and 30. It's up to you if you want to pick others; this can be picked randomly. The following code listing shows how this is done.

Listing 4.10   Creating hash functions

Define hash function (it is exactly like the createNum() function without the final conversion to a number and for 3 columns instead of 8).

Test if it works correctly (if no error is raised, it works). It's sampling on columns but all observations will be selected.

```
def hash_fn(x1,x2,x3):
    return [b'%d%d%d' % (i,j,k) for (i,j,k) in zip(x1,x2,x3)]

assert hash_fn([1,0],[1,1],[0,0]) == [b'110',b'010']

store['bucket1'] = hash_fn(data.movie10, data.movie15,data.movie28)
store['bucket2'] = hash_fn(data.movie7, data.movie18,data.movie22)
store['bucket3'] = hash_fn(data.movie16, data.movie19,data.movie30)
store.to_sql('movie_comparison',mc, flavor = 'mysql', index = True,
         index_label = 'cust_id', if_exists = 'replace')
```

Store this information in database.

Create hash values from customer movies, respectively [10,15, 28], [7,18, 22], [16,19, 30].

The hash function concatenates the values from the different movies into a binary value like what happened before in the createNum() function, only this time we don't convert to numbers and we only take 3 movies instead of 8 as input. The assert function shows how it concatenates the 3 values for every observation. When the client has bought movie 10 but not movies 15 and 28, it will return b'100' for bucket 1. When the client bought movies 7 and 18, but not 22, it will return b'110' for bucket 2. If we look at the current result we see the 4 variables we created earlier (bit1, bit2, bit3, bit4) from the 9 handpicked movies (figure 4.12).

	bit1	bit2	bit3	bit4	bucket1	bucket2	bucket3
0	10	62	42	182	011	100	011
1	23	28	223	180	001	111	001

Figure 4.12   Information from the bit string compression and the 9 sampled movies

The last trick we'll apply is indexing the customer table so lookups happen more quickly.

ADDING AN INDEX TO THE TABLE

Now you must add indices to speed up retrieval as needed in a real-time system. This is shown in the next listing.

Listing 4.11 Creating an index

> **Create function to easily create indices. Indices will quicken retrieval.**

```
def createIndex(column, cursor):
    sql = 'CREATE INDEX %s ON movie_comparison (%s);' % (column, column)
    cursor.execute(sql)

createIndex('bucket1',cursor)
createIndex('bucket2',cursor)
createIndex('bucket3',cursor)
```

> **Put index on bit buckets.**

With the data indexed we can now move on to the "model building part." In this case study no actual machine learning or statistical model is implemented. Instead we'll use a far simpler technique: string distance calculation. Two strings can be compared using the hamming distance as explained earlier in the theoretical intro to the case study.

4.5.4 Step 5: Model building

To use the hamming distance in the database we need to define it as a function.

CREATING THE HAMMING DISTANCE FUNCTION

We implement this as *a user-defined function*. This function can calculate the distance for a 32-bit integer (actually 4*8), as shown in the following listing.

Listing 4.12 Creating the hamming distance

```
Sql = '''
CREATE FUNCTION HAMMINGDISTANCE(
  A0 BIGINT, A1 BIGINT, A2 BIGINT, A3 BIGINT,
  B0 BIGINT, B1 BIGINT, B2 BIGINT, B3 BIGINT
)

RETURNS INT DETERMINISTIC
RETURN
    BIT_COUNT(A0 ^ B0) +
    BIT_COUNT(A1 ^ B1) +
    BIT_COUNT(A2 ^ B2) +
    BIT_COUNT(A3 ^ B3); '''

cursor.execute(Sql)

Sql = '''Select hammingdistance(
    b'11111111',b'00000000',b'11011111',b'11111111'
,b'11111111',b'10001001',b'11011111',b'11111111'
)'''
pd.read_sql(Sql,mc)
```

> **The function is stored in a database. You can only do this once; running this code a second time will result in an error: OperationalError: (1304, 'FUNCTION HAMMING-DISTANCE already exists').**

> **Define function. It takes 8 input arguments: 4 strings of length 8 for the first customer and another 4 strings of length 8 for the second customer. This way we can compare 2 customers side-by-side for 32 movies.**

> **To check this function you can run this SQL statement with 8 fixed strings. Notice the "b" before each string, indicating that you're passing bit values. The outcome of this particular test should be 3, which indicates the series of strings differ in only 3 places.**

> **This runs the query.**

If all is well, the output of this code should be 3.

Now that we have our hamming distance function in position, we can use it to find similar customers to a given customer, and this is exactly what we want our application to do. Let's move on to the last part: utilizing our setup as a sort of application.

4.5.5 *Step 6: Presentation and automation*

Now that we have it all set up, our application needs to perform two steps when confronted with a given customer:

- Look for similar customers.
- Suggest movies the customer has yet to see based on what he or she has already viewed and the viewing history of the similar customers.

First things first: select ourselves a lucky customer.

FINDING A SIMILAR CUSTOMER

Time to perform real-time queries. In the following listing, customer 27 is the happy one who'll get his next movies selected for him. But first we need to select customers with a similar viewing history.

Listing 4.13   Finding similar customers

We do two-step sampling. First sampling: index must be exactly the same as the one of the selected customer (is based on 9 movies). Selected people must have seen (or not seen) these 9 movies exactly like our customer did. Second sampling is a ranking based on the 4-bit strings. These take into account all the movies in the database.

Pick customer from database.

```
customer_id = 27
sql = "select * from movie_comparison where cust_id = %s" % customer_id
cust_data = pd.read_sql(sql,mc)
sql =  """ select cust_id,hammingdistance(bit1,
bit2,bit3,bit4,%s,%s,%s,%s) as distance
          from movie_comparison where bucket1 = '%s' or bucket2 ='%s'
or bucket3='%s' order by distance limit 3""" %
    (cust_data.bit1[0],cust_data.bit2[0],
cust_data.bit3[0], cust_data.bit4[0],
      cust_data.bucket1[0], cust_data.bucket2[0],cust_data.bucket3[0])
shortlist = pd.read_sql(sql,mc)
```

We show the 3 customers that most resemble customer 27. Customer 27 ends up first.

Table 4.5 shows customers 2 and 97 to be the most similar to customer 27. Don't forget that the data was generated randomly, so anyone replicating this example might receive different results.

Now we can finally select a movie for customer 27 to watch.

Table 4.5 The most similar customers to customer 27

	cust_id	distance
0	27	0
1	2	8
2	97	9

FINDING A NEW MOVIE

We need to look at movies customer 27 hasn't seen yet, but the nearest customer has, as shown in the following listing. This is also a good check to see if your distance function worked correctly. Although this may not be the closest customer, it's a good match with customer 27. By using the hashed indexes, you've gained enormous speed when querying large databases.

> **Listing 4.14 Finding an unseen movie**

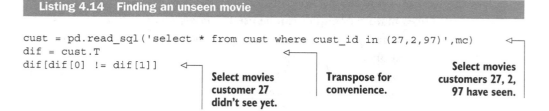

```
cust = pd.read_sql('select * from cust where cust_id in (27,2,97)',mc)
dif = cust.T
dif[dif[0] != dif[1]]
```

Select movies customer 27 didn't see yet.

Transpose for convenience.

Select movies customers 27, 2, 97 have seen.

Table 4.6 shows you can recommend movie 12, 15, or 31 based on customer 2's behavior.

Table 4.6 Movies from customer 2 can be used as suggestions for customer 27.

	0	1	2
Cust_id	2	27	97
Movie3	0	1	1
Movie9	0	1	1
Movie11	0	1	1
Movie12	1	0	0
Movie15	1	0	0
Movie16	0	1	1
Movie25	0	1	1
Movie31	1	0	0

Mission accomplished. Our happy movie addict can now indulge himself with a new movie, tailored to his preferences.

In the next chapter we'll look at even bigger data and see how we can handle that using the Horton Sandbox we downloaded in chapter 1.

4.6   Summary

This chapter discussed the following topics:

- The main *problems* you can run into when working with large data sets are these:
  - Not enough memory
  - Long-running programs
  - Resources that form bottlenecks and cause speed problems
- There are three main types of *solutions* to these problems:
  - Adapt your algorithms.
  - Use different data structures.
  - Rely on tools and libraries.
- Three main techniques can be used to *adapt an algorithm*:
  - Present algorithm data *one observation at a time* instead of loading the full data set at once.
  - *Divide matrices into smaller matrices* and use these to make your calculations.
  - Implement the *MapReduce* algorithm (using Python libraries such as Hadoopy, Octopy, Disco, or Dumbo).
- Three main *data structures* are used in data science. The first is a type of matrix that contains relatively little information, the *sparse matrix*. The second and third are data structures that enable you to retrieve information quickly in a large data set: the *hash function* and *tree structure*.
- Python has many *tools* that can help you deal with large data sets. Several tools will help you with the size of the volume, others will help you parallelize the computations, and still others overcome the relatively slow speed of Python itself. It's also easy to use Python as a tool to control other data science tools because Python is often chosen as a language in which to implement an API.
- The *best practices* from computer science are also valid in a data science context, so applying them can help you overcome the problems you face in a big data context.

First steps in big data

Over the last two chapters, we've steadily increased the size of the data. In chapter 3 we worked with data sets that could fit into the main memory of a computer. Chapter 4 introduced techniques to deal with data sets that were too large to fit in memory but could still be processed on a single computer. In this chapter you'll learn to work with technologies that can handle data that's so large a single node (computer) no longer suffices. In fact it may not even fit on a hundred computers. Now that's a challenge, isn't it?

We'll stay as close as possible to the way of working from the previous chapters; the focus is on giving you the confidence to work on a big data platform. To do this, the main part of this chapter is a case study. You'll create a dashboard that allows you to explore data from lenders of a bank. By the end of this chapter you'll have gone through the following steps:

- Load data into Hadoop, the most common big data platform.
- Transform and clean data with Spark.
- Store it into a big data database called Hive.
- Interactively visualize this data with Qlik Sense, a visualization tool.

All this (apart from the visualization) will be coordinated from within a Python script. The end result is a dashboard that allows you to explore the data, as shown in figure 5.1.

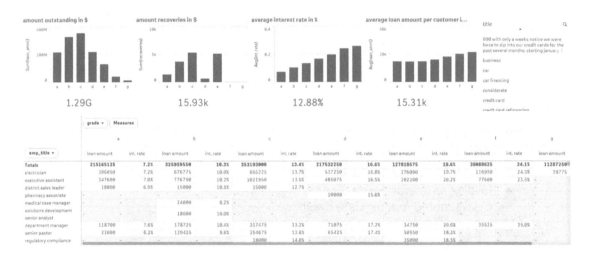

Figure 5.1   Interactive Qlik dashboard

Bear in mind that we'll only scratch the surface of both practice and theory in this introductory chapter on big data technologies. The case study will touch three big data technologies (Hadoop, Spark, and Hive), but only for data manipulation, not model building. It will be up to you to combine the big data technologies you get to see here with the model-building techniques we touched upon in previous chapters.

5.1 Distributing data storage and processing with frameworks

New big data technologies such as Hadoop and Spark make it much easier to work with and control a cluster of computers. Hadoop can scale up to thousands of computers, creating a cluster with petabytes of storage. This enables businesses to grasp the value of the massive amount of data available.

5.1.1 Hadoop: a framework for storing and processing large data sets

Apache Hadoop is a framework that simplifies working with a cluster of computers. It aims to be all of the following things and more:

- *Reliable*—By automatically creating multiple copies of the data and redeploying processing logic in case of failure.
- *Fault tolerant*—It detects faults and applies automatic recovery.
- *Scalable*—Data and its processing are distributed over clusters of computers (horizontal scaling).
- *Portable*—Installable on all kinds of hardware and operating systems.

The core framework is composed of a distributed file system, a resource manager, and a system to run distributed programs. In practice it allows you to work with the distributed file system almost as easily as with the local file system of your home computer. But in the background, the data can be scattered among thousands of servers.

THE DIFFERENT COMPONENTS OF HADOOP

At the heart of Hadoop we find

- A distributed file system (HDFS)
- A method to execute programs on a massive scale (MapReduce)
- A system to manage the cluster resources (YARN)

On top of that, an ecosystem of applications arose (figure 5.2), such as the databases Hive and HBase and frameworks for machine learning such as Mahout. We'll use Hive in this chapter. Hive has a language based on the widely used SQL to interact with data stored inside the database.

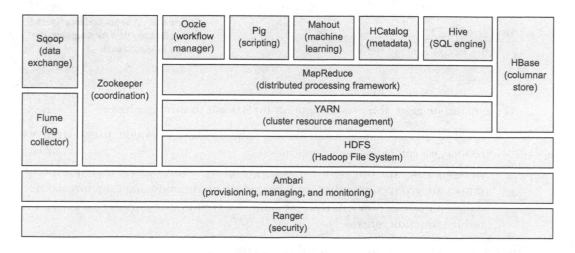

Figure 5.2 A sample from the ecosystem of applications that arose around the Hadoop Core Framework

It's possible to use the popular tool Impala to query Hive data up to 100 times faster. We won't go into Impala in this book, but more information can be found at http://impala.io/. We already had a short intro to MapReduce in chapter 4, but let's elaborate a bit here because it's such a vital part of Hadoop.

MAPREDUCE: HOW HADOOP ACHIEVES PARALLELISM

Hadoop uses a programming method called MapReduce to achieve parallelism. A MapReduce algorithm splits up the data, processes it in parallel, and then sorts, combines, and aggregates the results back together. However, the MapReduce algorithm isn't well suited for interactive analysis or iterative programs because it writes the data to a disk in between each computational step. This is expensive when working with large data sets.

Let's see how MapReduce would work on a small fictitious example. You're the director of a toy company. Every toy has two colors, and when a client orders a toy from the web page, the web page puts an order file on Hadoop with the colors of the toy. Your task is to find out how many color units you need to prepare. You'll use a MapReduce-style algorithm to count the colors. First let's look at a simplified version in figure 5.3.

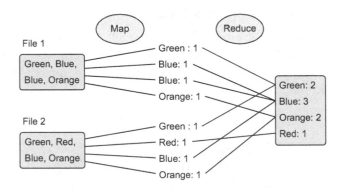

Figure 5.3   A simplified example of a MapReduce flow for counting the colors in input texts

As the name suggests, the process roughly boils down to two big phases:

- *Mapping phase*—The documents are split up into key-value pairs. Until we reduce, we can have many duplicates.
- *Reduce phase*—It's not unlike a SQL "group by." The different unique occurrences are grouped together, and depending on the reducing function, a different result can be created. Here we wanted a count per color, so that's what the reduce function returns.

In reality it's a bit more complicated than this though.

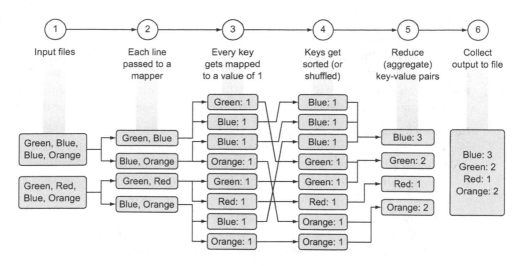

Figure 5.4   An example of a MapReduce flow for counting the colors in input texts

The whole process is described in the following six steps and depicted in figure 5.4.

1 Reading the input files.
2 Passing each line to a mapper job.
3 The mapper job parses the colors (keys) out of the file and outputs a file for each color with the number of times it has been encountered (value). Or more technically said, it maps a key (the color) to a value (the number of occurrences).
4 The keys get shuffled and sorted to facilitate the aggregation.
5 The reduce phase sums the number of occurrences per color and outputs one file per key with the total number of occurrences for each color.
6 The keys are collected in an output file.

NOTE   While Hadoop makes working with big data easy, setting up a good working cluster still isn't trivial, but cluster managers such as Apache Mesos do ease the burden. In reality, many (mid-sized) companies lack the competence to maintain a healthy Hadoop installation. This is why we'll work with the Hortonworks Sandbox, a pre-installed and configured Hadoop ecosystem. Installation instructions can be found in section 1.5: An introductory working example of Hadoop.

Now, keeping the workings of Hadoop in mind, let's look at Spark.

5.1.2   *Spark: replacing MapReduce for better performance*

Data scientists often do interactive analysis and rely on algorithms that are inherently iterative; it can take awhile until an algorithm converges to a solution. As this is a weak point of the MapReduce framework, we'll introduce the Spark Framework to overcome it. Spark improves the performance on such tasks by an order of magnitude.

WHAT IS SPARK?

Spark is a cluster computing framework similar to MapReduce. Spark, however, doesn't handle the storage of files on the (distributed) file system itself, nor does it handle the resource management. For this it relies on systems such as the Hadoop File System, YARN, or Apache Mesos. Hadoop and Spark are thus complementary systems. For testing and development, you can even run Spark on your local system.

HOW DOES SPARK SOLVE THE PROBLEMS OF MAPREDUCE?

While we oversimplify things a bit for the sake of clarity, Spark creates a kind of shared RAM memory between the computers of your cluster. This allows the different workers to share variables (and their state) and thus eliminates the need to write the intermediate results to disk. More technically and more correctly if you're into that: Spark uses Resilient Distributed Datasets (RDD), which are a distributed memory abstraction that lets programmers perform in-memory computations on large clusters in a fault-tolerant way.[1] Because it's an in-memory system, it avoids costly disk operations.

THE DIFFERENT COMPONENTS OF THE SPARK ECOSYSTEM

Spark core provides a NoSQL environment well suited for interactive, exploratory analysis. Spark can be run in batch and interactive mode and supports Python.

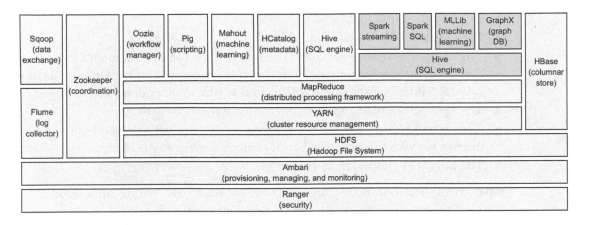

Figure 5.5　The Spark framework when used in combination with the Hadoop framework

Spark has four other large components, as listed below and depicted in figure 5.5.

1　Spark streaming is a tool for real-time analysis.
2　Spark SQL provides a SQL interface to work with Spark.
3　MLLib is a tool for machine learning inside the Spark framework.
4　GraphX is a graph database for Spark. We'll go deeper into graph databases in chapter 7.

---

[1]　See https://www.cs.berkeley.edu/~matei/papers/2012/nsdi_spark.pdf.

Now let's dip our toes into loan data using Hadoop, Hive, and Spark.

5.2   *Case study: Assessing risk when loaning money*

Enriched with a basic understanding of Hadoop and Spark, we're now ready to get our hands dirty on big data. The goal of this case study is to have a first experience with the technologies we introduced earlier in this chapter, and see that for a large part you can (but don't have to) work similarly as with other technologies. Note: The portion of the data used here isn't that big because that would require serious bandwidth to collect it and multiple nodes to follow along with the example.

What we'll use

- Horton Sandbox on a virtual machine. If you haven't downloaded and imported this to VM software such as VirtualBox, please go back to section 1.5 where this is explained. Version 2.3.2 of the Horton Sandbox was used when writing this chapter.
- Python libraries: Pandas and pywebhdsf. They don't need to be installed on your local virtual environment this time around; we need them directly on the Horton Sandbox. Therefore we need to fire up the Horton Sandbox (on VirtualBox, for instance) and make a few preparations.

In the Sandbox command line there are several things you still need to do for this all to work, so connect to the command line. You can do this using a program like PuTTY. If you're unfamiliar with PuTTY, it offers a command line interface to servers and can be downloaded freely at http://www.chiark.greenend.org.uk/~sgtatham/putty/download.html.

The PuTTY login configuration is shown in figure 5.6.

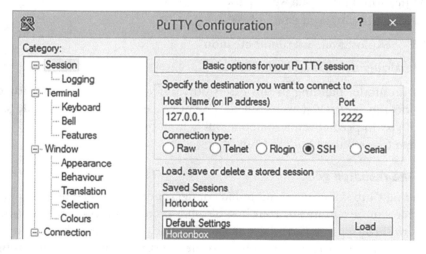

Figure 5.6   Connecting to Horton Sandbox using PuTTY

The default user and password are (at the time of writing) "root" and "hadoop", respectively. You'll need to change this password at the first login, though.

Once connected, issue the following commands:

- `yum -y install python-pip`—This installs pip, a Python package manager.
- `pip install git+https://github.com/DavyCielen/pywebhdfs.git -upgrade`—At the time of writing there was a problem with the pywebhdfs library and we fixed that in this fork. Hopefully you won't require this anymore when you read this; the problem has been signaled and should be resolved by the maintainers of this package.
- `pip install pandas`—To install Pandas. This usually takes awhile because of the dependencies.

An .ipynb file is available for you to open in Jupyter or (the older) Ipython and follow along with the code in this chapter. Setup instructions for Horton Sandbox are repeated there; make sure to run the code directly on the Horton Sandbox. Now, with the preparatory business out of the way, let's look at what we'll need to do.

In this exercise, we'll go through several more of the data science process steps:

Step 1: The research goal. This consists of two parts:

- Providing our manager with a dashboard
- Preparing data for other people to create their own dashboards

Step 2: Data retrieval

- Downloading the data from the lending club website
- Putting the data on the Hadoop File System of the Horton Sandbox

Step 3: Data preparation

- Transforming this data with Spark
- Storing the prepared data in Hive

Steps 4 & 6: Exploration and report creation

- Visualizing the data with Qlik Sense

We have no model building in this case study, but you'll have the infrastructure in place to do this yourself if you want to. For instance, you can use SPARK Machine learning to try to predict when someone will default on his debt.

It's time to meet the Lending Club.

5.2.1   *Step 1: The research goal*

The Lending Club is an organization that connects people in need of a loan with people who have money to invest. Your boss also has money to invest and wants information before throwing a substantial sum on the table. To achieve this, you'll create a report for him that gives him insight into the average rating, risks, and return for lending money to a certain person. By going through this process, you make the data accessible in a dashboard tool, thus enabling other people to explore it as well. In a sense this is the secondary goal of this case: opening up the data for self-service BI. Self-service Business Intelligence is often applied in data-driven organizations that don't have analysts to spare. Anyone in the organization can do the simple slicing and dicing themselves while leaving the more complicated analytics for the data scientist.

We can do this case study because the Lending Club makes anonymous data available about the existing loans. By the end of this case study, you'll create a report similar to figure 5.7.

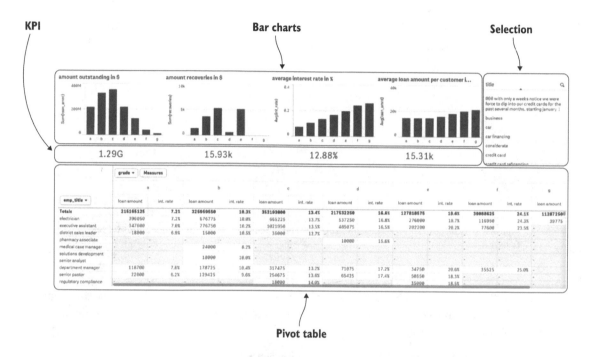

Figure 5.7 The end result of this exercise is an explanatory dashboard to compare a lending opportunity to similar opportunities.

First things first, however; let's get ourselves data.

5.2.2 Step 2: Data retrieval

It's time to work with the Hadoop File System (or hdfs). First we'll send commands through the command line and then through the Python scripting language with the help of the pywebhdfs package.

The Hadoop file system is similar to a normal file system, except that the files and folders are stored over multiple servers and you don't know the physical address of each file. This is not unfamiliar if you've worked with tools such as Dropbox or Google Drive. The files you put on these drives are stored somewhere on a server without you knowing exactly on which server. As on a normal file system, you can create, rename, and delete files and folders.

USING THE COMMAND LINE TO INTERACT WITH THE HADOOP FILE SYSTEM

Let's first retrieve the currently present list of directories and files in the Hadoop root folder using the command line. Type the command hadoop fs -ls / in PuTTY to achieve this.

Make sure you turn on your virtual machine with the Hortonworks Sandbox before attempting a connection. In PuTTY you should then connect to 127.0.0.1:2222, as shown before in figure 5.6.

The output of the Hadoop command is shown in figure 5.8. You can also add arguments such as hadoop fs -ls -R / to get a recursive list of all the files and subdirectories.

```
[root@sandbox ~]# hadoop fs -ls /
Found 20 items
drwxrwxrwx   - admin  hadoop          0 2015-07-14 14:54 /LoanStats3c.cs
-rw-r--r--   1 root   hadoop  120834552 2015-07-14 14:47 /LoanStats3c.csv
drwxrwxrwx   - yarn   hadoop          0 2015-07-15 13:32 /app-logs
drwxr-xr-x   - hdfs   hdfs            0 2015-06-05 09:19 /apps
drwxr-xr-x   - admin  hadoop          0 2015-07-13 06:47 /book
drwxr-xr-x   - root   hadoop          0 2015-07-17 10:24 /chapter5
-rwxr-xr-x   1 hdfs   hadoop       4240 2015-07-14 19:32 /cout.json
```

Figure 5.8   Output from the Hadoop list command: hadoop fs -ls /. The Hadoop root folder is listed.

We'll now create a new directory "chapter5" on hdfs to work with during this chapter. The following commands will create the new directory and give everybody access to the folder:

```
sudo -u hdfs hadoop fs -mkdir /chapter5
sudo -u hdfs hadoop fs -chmod 777 /chapter5
```

You probably noticed a pattern here. The Hadoop commands are very similar to our local file system commands (POSIX style) but start with Hadoop fs and have a dash - before each command. Table 5.1 gives an overview of popular file system commands on Hadoop and their local file system command counterparts.

Table 5.1   List of common Hadoop file system commands

Goal	Hadoop file system command	Local file system command
Get a list of files and directories from a directory	hadoop fs -ls URI	ls URI
Create a directory	hadoop fs -mkdir URI	mkdir URI
Remove a directory	hadoop fs -rm -r URI	rm -r URI

Table 5.1 List of common Hadoop file system commands

Goal	Hadoop file system command	Local file system command
Change the permission of files	hadoop fs –chmod MODE URI	chmod MODE URI
Move or rename file	hadoop fs –mv OLDURI NEWURI	mv OLDURI NEWURI

There are two special commands you'll use often. These are

- Upload files from the local file system to the distributed file system (`hadoop fs -put LOCALURI REMOTEURI`).
- Download a file from the distributed file system to the local file system (`hadoop -get REMOTEURI`).

Let's clarify this with an example. Suppose you have a .CSV file on the Linux virtual machine from which you connect to the Linux Hadoop cluster. You want to copy the .CSV file from your Linux virtual machine to the cluster hdfs. Use the command `hadoop -put mycsv.csv /data`.

Using PuTTY we can start a Python session on the Horton Sandbox to retrieve our data using a Python script. Issue the "pyspark" command in the command line to start the session. If all is well you should see the welcome screen shown in figure 5.9.

```
Welcome to

      __         __
     / __/__  ___ _____/ /__
    _\ \/ _ \/ _ `/ __/  '_/
   /__ / .__/\_,_/_/ /_/\_\   version 1.4.1
      /_/

Using Python version 2.6.6 (r266:84292, Jul 23 2015 15:22:56)
SparkContext available as sc, HiveContext available as sqlContext.
>>>
```

Figure 5.9 The welcome screen of Spark for interactive use with Python

Now we use Python code to fetch the data for us, as shown in the following listing.

Listing 5.1 Drawing in the Lending Club loan data

```
import requests
import zipfile
import StringIO
source = requests.get("https://resources.lendingclub.com/
    LoanStats3d.csv.zip", verify=False)
stringio = StringIO.StringIO(source.content)
unzipped = zipfile.ZipFile(stringio)
```

> **Downloads data from Lending Club. This is https so it should verify, but we won't bother (verify=False).**

> **Creates virtual file.**

> **Unzips data.**

We download the file "LoanStats3d.csv.zip" from the Lending Club's website at https://resources.lendingclub.com/LoanStats3d.csv.zip and unzip it. We use methods from the requests, zipfile, and stringio Python packages to respectively download the data, create a virtual file, and unzip it. This is only a single file; if you want all their data you could create a loop, but for demonstration purposes this will do. As we mentioned before, an important part of this case study will be data preparation with big data technologies. Before we can do so, however, we need to put it on the Hadoop file system. PyWebHdfs is a package that allows you to interact with the Hadoop file system from Python. It translates and passes your commands to rest calls for the webhdfs interface. This is useful because you can use your favorite scripting language to automate tasks, as shown in the following listing.

> **Listing 5.2   Storing data on Hadoop**

Stores it locally because we need to transfer it to Hadoop file system.

Does preliminary data cleaning using Pandas: removes top row and bottom 2 rows because they're useless. Opening original file will show you this.

```
import pandas as pd
from pywebhdfs.webhdfs import PyWebHdfsClient
subselection_csv = pd.read_csv(unzipped.open('LoanStats3d.csv'),
        skiprows=1,skipfooter=2,engine='python')
stored_csv = subselection_csv.to_csv('./stored_csv.csv')
hdfs = PyWebHdfsClient(user_name="hdfs",port=50070,host="sandbox")
hdfs.make_dir('chapter5')
with open('./stored_csv.csv') as file_data:
        hdfs.create_file('chapter5/LoanStats3d.csv',file_data,
        overwrite=True)
```

Creates folder "chapter5" on Hadoop file system.

Connects to Hadoop Sandbox.

Creates .csv file on Hadoop file system.

Opens locally stored csv.

We had already downloaded and unzipped the file in listing 5.1; now in listing 5.2 we made a sub-selection of the data using Pandas and stored it locally. Then we created a directory on Hadoop and transferred the local file to Hadoop. The downloaded data is in .CSV format and because it's rather small, we can use the Pandas library to remove the first line and last two lines from the file. These contain comments and will only make working with this file cumbersome in a Hadoop environment. The first line of our code imports the Pandas package, while the second line parses the file into memory and removes the first and last two data lines. The third code line saves the data to the local file system for later use and easy inspection.

Before moving on, we can check our file using the following line of code:

```
print hdfs.get_file_dir_status('chapter5/LoanStats3d.csv')
```

The PySpark console should tell us our file is safe and well on the Hadoop system, as shown in figure 5.10.

```
>>> print hdfs.get_file_dir_status('chapter5/LoanStats3d.csv')#A
{u'FileStatus': {u'group': u'hdfs', u'permission': u'755', u'blockSize': 1342177
28, u'accessTime': 1449236321223, u'pathSuffix': u'', u'modificationTime': 14492
36321965, u'replication': 3, u'length': 120997124, u'childrenNum': 0, u'owner':
u'hdfs', u'storagePolicy': 0, u'type': u'FILE', u'fileId': 17520}}
```

Figure 5.10   Retrieve file status on Hadoop via the PySpark console

With the file ready and waiting for us on Hadoop, we can move on to data preparation using Spark, because it's not clean enough to directly store in Hive.

5.2.3   *Step 3: Data preparation*

Now that we've downloaded the data for analysis, we'll use Spark to clean the data before we store it in Hive.

DATA PREPARATION IN SPARK

Cleaning data is often an interactive exercise, because you spot a problem and fix the problem, and you'll likely do this a couple of times before you have clean and crisp data. An example of dirty data would be a string such as "UsA", which is improperly capitalized. At this point, we no longer work in jobs.py but use the PySpark command line interface to interact directly with Spark.

Spark is well suited for this type of interactive analysis because it doesn't need to save the data after each step and has a much better model than Hadoop for sharing data between servers (a kind of distributed memory).

The transformation consists of four parts:

1 Start up PySpark (should still be open from section 5.2.2) and load the Spark and Hive context.
2 Read and parse the .CSV file.
3 Split the header line from the data.
4 Clean the data.

Okay, onto business. The following listing shows the code implementation in the PySpark console.

Listing 5.3   Connecting to Apache Spark

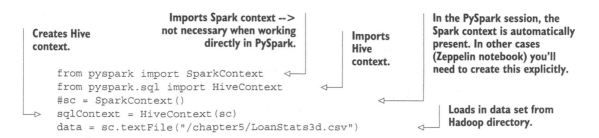

```
from pyspark import SparkContext
from pyspark.sql import HiveContext
#sc = SparkContext()
sqlContext = HiveContext(sc)
data = sc.textFile("/chapter5/LoanStats3d.csv")
```

Creates Hive context.

Imports Spark context --> not necessary when working directly in PySpark.

Imports Hive context.

In the PySpark session, the Spark context is automatically present. In other cases (Zeppelin notebook) you'll need to create this explicitly.

Loads in data set from Hadoop directory.

```
parts = data.map(lambda r:r.split(','))
firstline = parts.first()
datalines = parts.filter(lambda x:x != firstline)
def cleans(row):
        row [7] = str(float(row [7][:-1])/100)
        return [s.encode('utf8').replace(r"_"," ").lower() for s in row]
datalines = datalines.map(lambda x: cleans(x))
```

Column 8 (index = 7) has % formatted numbers. We don't need that % sign.

Executes data cleaning line by line.

Encodes everything in utf8, replaces underscores with spaces, and lowercases everything.

Grabs first line.

Cleaning function will use power of Spark to clean data. The input of this function will be a line of data.

Splits data set with comma (,) delimiter. This is the end of line delimiter for this file.

Grabs all lines but first line, because first line is only variable names.

Let's dive a little further into the details for each step.

Step 1: Starting up Spark in interactive mode and loading the context

The Spark context import isn't required in the PySpark console because a context is readily available as variable sc. You might have noticed this is also mentioned when opening PySpark; check out figure 5.9 in case you overlooked it. We then load a Hive context to enable us to work interactively with Hive. If you work interactively with Spark, the Spark and Hive contexts are loaded automatically, but if you want to use it in batch mode you need to load it manually. To submit the code in batch you would use the spark-submit filename.py command on the Horton Sandbox command line.

```
from pyspark import SparkContext
from pyspark.sql import HiveContext
sc = SparkContext()
sqlContext = HiveContext(sc)
```

With the environment set up, we're ready to start parsing the .CSV file.

Step 2: Reading and parsing the .CSV file

Next we read the file from the Hadoop file system and split it at every comma we encounter. In our code the first line reads the .CSV file from the Hadoop file system. The second line splits every line when it encounters a comma. Our .CSV parser is naïve by design because we're learning about Spark, but you can also use the .CSV package to help you parse a line more correctly.

```
data = sc.textFile("/chapter5/LoanStats3d.csv")
parts = data.map(lambda r:r.split(','))
```

Notice how similar this is to a functional programming approach. For those who've never encountered it, you can naïvely read `lambda r:r.split(',')` as "for every input r (a row in this case), split this input r when it encounters a comma." As in this case, "for every input" means "for every row," but you can also read it as "split every row by a comma." This functional-like syntax is one of my favorite characteristics of Spark.

Step 3: Split the header line from the data

To separate the header from the data, we read in the first line and retain every line that's not similar to the header line:

```
firstline = parts.first()
datalines = parts.filter(lambda x:x != firstline)
```

Following the best practices in big data, we wouldn't have to do this step because the first line would already be stored in a separate file. In reality, .CSV files do often contain a header line and you'll need to perform a similar operation before you can start cleaning the data.

Step 4: Clean the data

In this step we perform basic cleaning to enhance the data quality. This allows us to build a better report.

After the second step, our data consists of arrays. We'll treat every input for a lambda function as an array now and return an array. To ease this task, we build a helper function that cleans. Our cleaning consists of reformatting an input such as "10,4%" to 0.104 and encoding every string as utf-8, as well as replacing underscores with spaces and lowercasing all the strings. The second line of code calls our helper function for every line of the array.

```
def cleans(row):
        row [7] = str(float(row [7][:-1])/100)
        return [s.encode('utf8').replace(r"_"," ").lower() for s in row]
datalines = datalines.map(lambda x: cleans(x))
```

Our data is now prepared for the report, so we need to make it available for our reporting tools. Hive is well suited for this, because many reporting tools can connect to it. Let's look at how to accomplish this.

SAVE THE DATA IN HIVE

To store data in Hive we need to complete two steps:

1 Create and register metadata.
2 Execute SQL statements to save data in Hive.

In this section, we'll once again execute the next piece of code in our beloved PySpark shell, as shown in the following listing.

Listing 5.4   Storing data in Hive (full)

Creates metadata: the Spark SQL StructField function represents a field
in a StructType. The StructField object is comprised of three fields:
name (a string), dataType (a DataType), and "nullable" (a boolean). The
field of name is the name of a StructField. The field of dataType specifies
the data type of a StructField. The field of nullable specifies if values of a
StructField can contain None values.

Imports SQL
data types.

```
from pyspark.sql.types import *
fields = [StructField(field_name,StringType(),True) for field_name in
    firstline]
schema = StructType(fields)
schemaLoans = sqlContext.createDataFrame(datalines, schema)
schemaLoans.registerTempTable("loans")
```

Creates data
frame from data
(datalines) and
data schema
(schema).

Registers it
as a table
called loans.

**StructType function creates
the data schema. A StructType
object requires a list of
StructFields as input.**

```
sqlContext.sql("drop table if exists LoansByTitle")
sql = '''create table LoansByTitle stored as parquet as select title,
    count(1) as number from loans group by title order by number desc'''
sqlContext.sql(sql)

sqlContext.sql('drop table if exists raw')
sql = '''create table raw stored as parquet as select title,
    emp_title,grade,home_ownership,int_rate,recoveries,
    collection_recovery_fee,loan_amnt,term from loans'''
```

**Drops table (in case
it already exists) and
stores a subset of
raw data in Hive.**

**Drops table (in case it already exists), summarizes,
and stores it in Hive. LoansByTitle represents the
sum of loans by job title.**

Let's drill deeper into each step for a bit more clarification.

Step 1: Create and register metadata

Many people prefer to use SQL when they work with data. This is also possible with
Spark. You can even read and store data in Hive directly as we'll do. Before you can do
that, however, you'll need to create metadata that contains a column name and column type for every column.

The first line of code is the imports. The second line parses the field name and the
field type and specifies if a field is mandatory. The `StructType` represents rows as an
array of structfields. Then you place it in a dataframe that's registered as a (temporary) table in Hive.

```
from pyspark.sql.types import *
fields = [StructField(field_name,StringType(),True) for field_name in firstline]
schema = StructType(fields)
schemaLoans = sqlContext.createDataFrame(datalines, schema)
schemaLoans.registerTempTable("loans")
```

With the metadata ready, we're now able to insert the data into Hive.

Step 2: Execute queries and store table in Hive

Now we're ready to use a SQL-dialect on our data. First we'll make a summary table that counts the number of loans per purpose. Then we store a subset of the cleaned raw data in Hive for visualization in Qlik.

Executing SQL-like commands is as easy as passing a string that contains the SQL-command to the `sqlContext.sql` function. Notice that we aren't writing pure SQL because we're communicating directly with Hive. Hive has its own SQL-dialect called HiveQL. In our SQL, for instance, we immediately tell it to store the data as a Parquet file. Parquet is a popular big data file format.

```
sqlContext.sql("drop table if exists LoansByTitle")
sql = '''create table LoansByTitle stored as parquet as select title,
    count(1) as number from loans group by title order by number desc'''
sqlContext.sql(sql)

sqlContext.sql('drop table if exists raw')
sql = '''create table raw stored as parquet as select title,
    emp_title,grade,home_ownership,int_rate,recoveries,collection_recovery_f
    ee,loan_amnt,term from loans'''
sqlContext.sql(sql)
```

With the data stored in Hive, we can connect our visualization tools to it.

5.2.4 Step 4: Data exploration & Step 6: Report building

We'll build an interactive report with Qlik Sense to show to our manager. Qlik Sense can be downloaded from http://www.qlik.com/try-or-buy/download-qlik-sense after subscribing to their website. When the download begins you will be redirected to a page containing several informational videos on how to install and work with Qlik Sense. It's recommended to watch these first.

We use the Hive ODBC connector to read data from Hive and make it available for Qlik. A tutorial on installing ODBC connectors in Qlik is available. For major operating systems, this can be found at http://hortonworks.com/hdp/addons/.

> **NOTE** In Windows, this might not work out of the box. Once you install the ODBC, make sure to check your Windows ODBC manager (CTRL+F and look for ODBC). In the manager, go to "System-DSN" and select the "Sample Hive Hortonworks DSN". Make sure your settings are correct (as shown in figure 5.11) or Qlik won't connect to the Hortonworks Sandbox.

Figure 5.11   Windows Hortonworks ODBC configuration

Let's hope you didn't forget your Sandbox password; as you can see in figure 5.11, you need it again.

Now open Qlik Sense. If installed in Windows you should have gotten the option to place a shortcut to the .exe on your desktop. Qlik isn't freeware; it's a commercial product with a bait version for single customers, but it will suffice for now. In the last chapter we'll create a dashboard using free JavaScript libraries.

Qlik can either take the data directly into memory or make a call every time to Hive. We've chosen the first method because it works faster.

This part has three steps:

1 Load data inside Qlik with an ODBC connection.
2 Create the report.
3 Explore data.

Let start with the first step, loading data into Qlik.

Step 1: Load data in Qlik
When you start Qlik Sense it will show you a welcome screen with the existing reports (called apps), as shown in figure 5.12.

Figure 5.12   The Qlik Sense welcome screen

To start a new app, click on the *Create new app* button on the right of the screen, as shown in figure 5.13. This opens up a new dialog box. Enter "chapter 5" as the new name of our app.

Figure 5.13   The Create new app message box

A confirmation box appears (figure 5.14) if the app is created successfully.

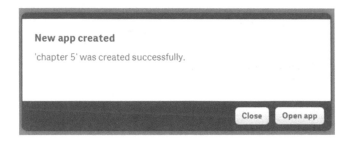

New app created

'chapter 5' was created successfully.

Close   Open app

Figure 5.14   A box confirms that the app was created successfully.

Click on the Open app button and a new screen will prompt you to add data to the application (figure 5.15).

Get started adding data to your app.

Add data
Add data from a file, a
database or Qlik DataMarket.

Data load editor
Load data from files or
databases, and perform data
transformation with the data
load script.

Figure 5.15   A start-adding-data screen pops up when you open a new app.

Click on the Add data button and choose ODBC as a data source (figure 5.16).

In the next screen (figure 5.17) select User DSN, Hortonworks, and specify the root as username and hadoop as a password (or the new one you gave when logging into the Sandbox for the first time).

NOTE   The Hortonworks option doesn't show up by default. You need to install the HDP 2.3 ODBC connector for this option to appear (as stated before). If you haven't succeeded in installing it at this point, clear instructions for this can be found at https://blogs.perficient.com/multi-shoring/blog/2015/09/29/how-to-connect-hortonworks-hive-from-qlikview-with-odbc-driver/.

Click on the arrow pointing to the right to go to the next screen.

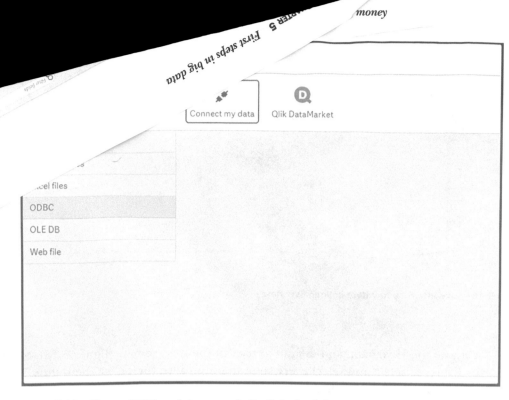

Figure 5.16 Choose ODBC as data source in the Select a data source screen

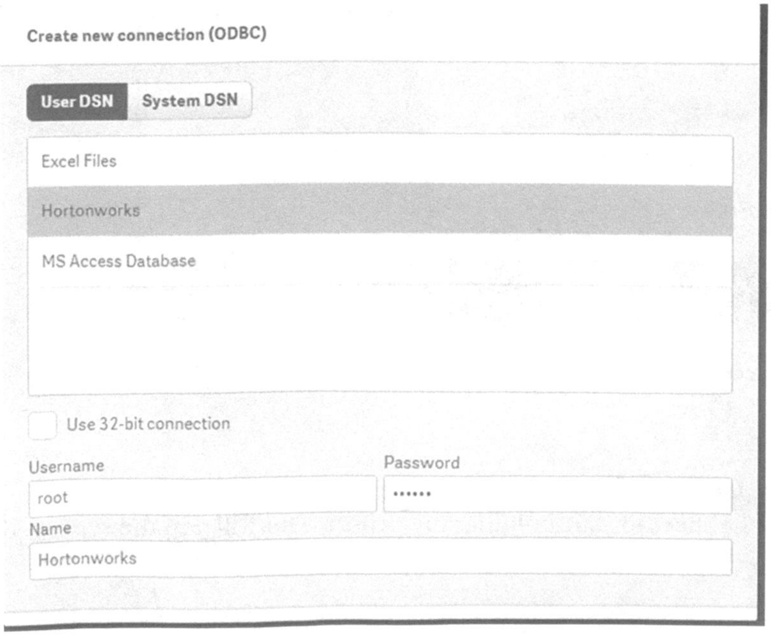

Figure 5.17 Choose Hortonworks on the User DSN and specify the username and password.

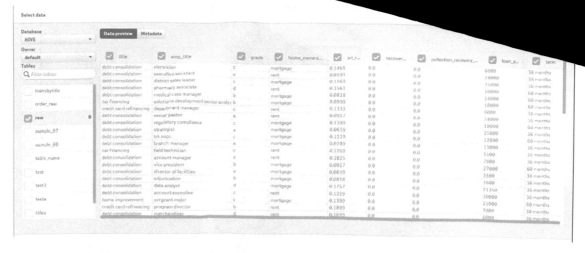

Figure 5.18 Hive interface raw data column overview

Choose the Hive data, and default as user in the next screen (figure 5.18). Select raw as Tables to select and select every column for import; then click the button Load and Finish to complete this step.

After this step, it will take a few seconds to load the data in Qlik (figure 5.19).

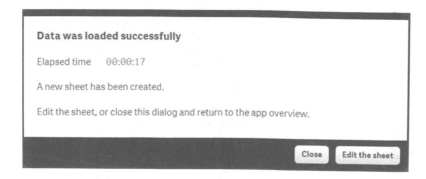

Figure 5.19 A confirmation that the data is loaded in Qlik

Step 2: Create the report
Choose Edit the sheet to start building the report. This will add the report editor (figure 5.20).

Figure 5.20 An editor screen for reports opens

Substep 1: Adding a selection filter to the report   The first thing we'll add to the report is a selection box that shows us why each person wants a loan. To achieve this, drop the title measure from the left asset panel on the report pane and give it a comfortable size and position (figure 5.21). Click on the Fields table so you can drag and drop fields.

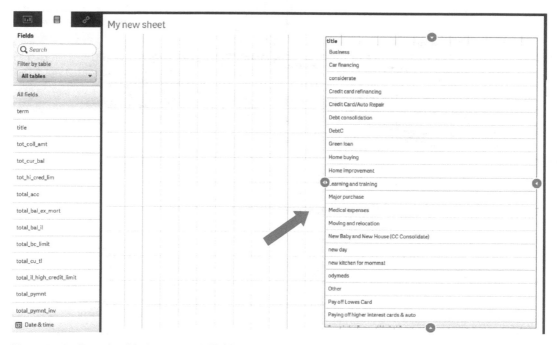

Figure 5.21 Drag the title from the left Fields pane to the report pane.

Substep 2: Adding a KPI to the report   A KPI chart shows an aggregated number for the total population that's selected. Numbers such as the average interest rate and the total number of customers are shown in this chart (figure 5.22).

Adding a KPI to a report takes four steps, as listed below and shown in figure 5.23.

Average Interest Rate

0.13

Figure 5.22   An example of a KPI chart

1   *Choose a chart*—Choose KPI as the chart and place it on the report screen; resize and position to your liking.
2   *Add a measure*—Click the add measure button inside the chart and select int_rate.
3   *Choose an aggregation method*—Avg(int_rate).
4   *Format the chart*—On the right pane, fill in average interest rate as Label.

Figure 5.23   The four steps to add a KPI chart to a Qlik report

In total we'll add four KPI charts to our report, so you'll need to repeat these steps for the following KPI's:

- Average interest rate
- Total loan amount
- Average loan amount
- Total recoveries

Substep 3: Adding bar charts to our report Next we'll add four bar charts to the report. These will show the different numbers for each risk grade. One bar chart will explain the average interest rate per risk group, and another will show us the total loan amount per risk group (figure 5.24).

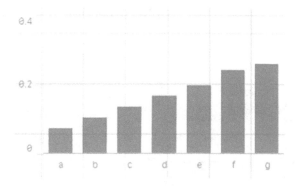

Figure 5.24 An example of a bar chart

Adding a bar chart to a report takes five steps, as listed below and shown in figure 5.25.

1. *Choose a chart*—Choose bar chart as the chart and place it on the report screen; resize and position to your liking.
2. *Add a measure*—Click the Add measure button inside the chart and select int_rate.
3. *Choose an aggregation method*—Avg(int_rate).
4. *Add a dimension*—Click Add dimension, and choose grade as the dimension.
5. *Format the chart*—On the right pane, fill in average interest rate as Label.

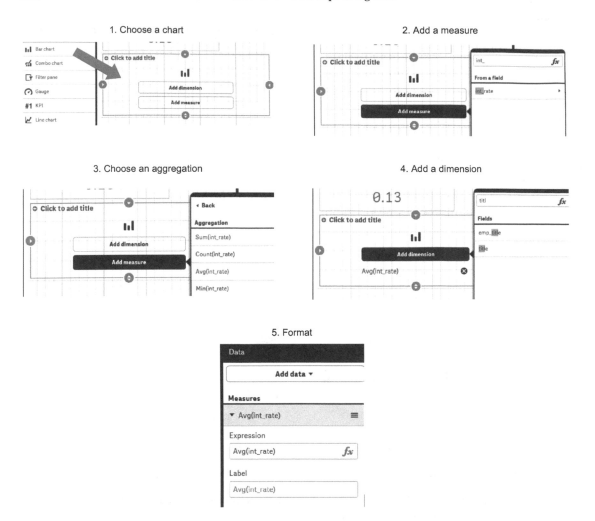

Figure 5.25   Adding a bar chart takes five steps.

Repeat this procedure for the following dimension and measure combinations:

- Average interest rate per grade
- Average loan amount per grade
- Total loan amount per grade
- Total recoveries per grade

Substep 4: Adding a cross table to the report   Suppose you want to know the average interest rate paid by directors of risk group C. In this case you want to get a measure (interest rate) for a combination of two dimensions (job title and risk grade). This can be achieved with a pivot table such as in figure 5.26.

average interest rate per job title / risk grade

emp_title ▼	grade ▼			
	a	b	c	d
electrician	0.072455172	0.099825	0.13674545	0.16769333
executive assistant	0.069928571	0.1023193	0.13515811	0.16489091
district sales leader	0.0692	0.1049	0.1269	-
pharmacy associate	-	-	-	0.1561
medical case manager	-	0.0818	-	-
solutions development senior analyst	-	0.0999	-	-
department manager	0.075866667	0.10396667	0.132172	0.17185

Figure 5.26   An example of a pivot table, showing the average interest rate paid per job title/risk grade combination

Adding a pivot table to a report takes six steps, as listed below and shown in figure 5.27.

1 *Choose a chart*—Choose pivot table as the chart and place it on the report screen; resize and position to your liking.
2 *Add a measure*—Click the Add measure button inside the chart and select int_rate.
3 *Choose an aggregation method*—Avg(int_rate).
4 *Add a row dimension*—Click Add dimension, and choose emp_title as the dimension.
5 *Add a column dimension*—Click Add data, choose column, and select grade.
6 *Format the chart*—On the right pane, fill in average interest rate as Label.

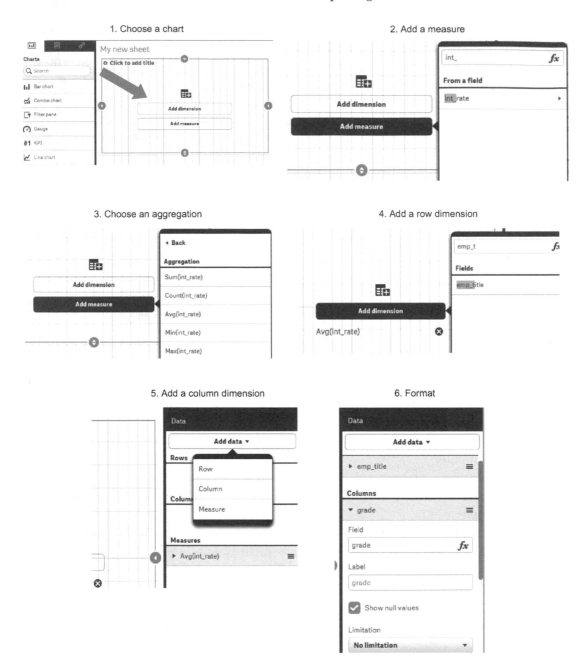

Figure 5.27   Adding a pivot table takes six steps.

Figure 5.28   The end result in edit mode

After resizing and repositioning, you should achieve a result similar to figure 5.28. Click the Done button on the left and you're ready to explore the data.

Step 3: Explore the data

The result is an interactive graph that updates itself based on the selections you make. Why don't you try to look for the information from directors and compare them to artists? To achieve this, hit the emp_title in the pivot table and type director in the search field. The result looks like figure 5.29. In the same manner, we can look at the artists, as shown in figure 5.30. Another interesting insight comes from comparing the rating for home-buying purposes with debt consolidation purposes.

We finally did it: We created the report our manager craves, and in the process we opened the door for other people to create their own reports using this data. An interesting next step for you to ponder on would be to use this setup to find those people likely to default on their debt. For this you can use the Spark Machine learning capabilities driven by online algorithms like the ones demonstrated in chapter 4.

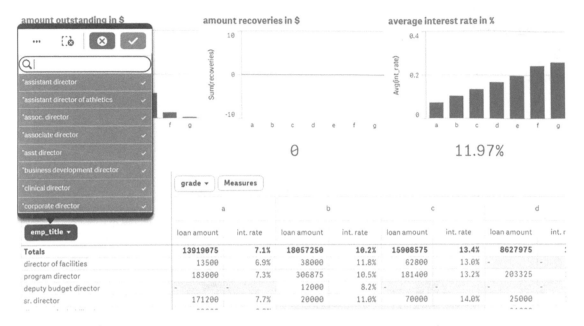

Figure 5.29   When we select directors, we can see that they pay an average rate of 11.97% for a loan.

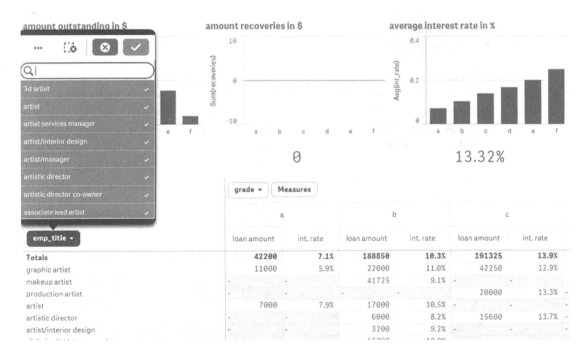

Figure 5.30   When we select artists, we see that they pay an average interest rate of 13.32% for a loan.

In this chapter we got a hands-on introduction to the Hadoop and Spark frameworks. We covered a lot of ground, but be honest, Python makes working with big data technologies dead easy. In the next chapter we'll dig deeper into the world of NoSQL databases and come into contact with more big data technologies.

5.3   *Summary*

In this chapter you learned that

- Hadoop is a framework that enables you to store files and distribute calculations amongst many computers.
- Hadoop hides all the complexities of working with a cluster of computers for you.
- An ecosystem of applications surrounds Hadoop and Spark, ranging from databases to access control.
- Spark adds a shared memory structure to the Hadoop Framework that's better suited for data science work.
- In the chapter case study we used PySpark (a Python library) to communicate with Hive and Spark from Python. We used the pywebhdfs Python library to work with the Hadoop library, but you could do as well using the OS command line.
- It's easy to connect a BI tool such as Qlik to Hadoop.

Join the NoSQL movement

This chapter covers

- Understanding NoSQL databases and why they're used today
- Identifying the differences between NoSQL and relational databases
- Defining the ACID principle and how it relates to the NoSQL BASE principle
- Learning why the CAP theorem is important for multi-node database setup
- Applying the data science process to a project with the NoSQL database Elasticsearch

This chapter is divided into two parts: a theoretical start and a practical finish.

- In the first part of this chapter we'll look into NoSQL databases in general and answer these questions: Why do they exist? Why not until recently? What types are there and why should you care?
- In part two we'll tackle a real-life problem—disease diagnostics and profiling—using freely available data, Python, and a NoSQL database.

No doubt you've heard about NoSQL databases and how they're used religiously by many high-tech companies. But what are NoSQL databases and what makes them so different from the relational or SQL databases you're used to? *NoSQL* is short for *Not Only Structured Query Language*, but although it's true that NoSQL databases can allow you to query them with SQL, you don't have to focus on the actual name. Much debate has already raged over the name and whether this group of new databases should even have a collective name at all. Rather, let's look at what they represent as opposed to *relational database management systems (RDBMS)*. Traditional databases reside on a single computer or server. This used to be fine as a long as your data didn't outgrow your server, but it hasn't been the case for many companies for a long time now. With the growth of the internet, companies such as Google and Amazon felt they were held back by these single-node databases and looked for alternatives.

Numerous companies use single-node NoSQL databases such as MongoDB because they want the flexible schema or the ability to hierarchically aggregate data. Here are several early examples:

- Google's first NoSQL solution was Google BigTable, which marked the start of the *columnar databases.*[1]
- Amazon came up with Dynamo, *a key-value store*.[2]
- Two more database types emerged in the quest for partitioning: the *document store* and the *graph database*.

We'll go into detail on each of the four types later in the chapter.

Please note that, although size was an important factor, these databases didn't originate solely from the need to handle larger volumes of data. Every *V* of big data has influence (volume, variety, velocity, and sometimes veracity). Graph databases, for instance, can handle network data. Graph database enthusiasts even claim that everything can be seen as a network. For example, how do you prepare dinner? With ingredients. These ingredients are brought together to form the dish and can be used along with other ingredients to form other dishes. Seen from this point of a view, ingredients and recipes are part of a network. But recipes and ingredients could also be stored in your relational database or a document store; it's all how you look at the problem. Herein lies the strength of NoSQL: the ability to look at a problem from a different angle, shaping the data structure to the use case. As a data scientist, your job is to find the best answer to any problem. Although sometimes this is still easier to attain using RDBMS, often a particular NoSQL database offers a better approach.

Are relational databases doomed to disappear in companies with big data because of the need for partitioning? No, NewSQL platforms (not to be confused with NoSQL) are the RDBMS answer to the need for cluster setup. NewSQL databases follow the relational model but are capable of being divided into a distributed cluster like NoSQL

---

[1] See http://static.googleusercontent.com/media/research.google.com/en//archive/bigtable-osdi06.pdf.
[2] See http://www.allthingsdistributed.com/files/amazon-dynamo-sosp2007.pdf.

databases. It's not the end of relational databases and certainly not the end of SQL, as platforms like Hive translate SQL into MapReduce jobs for Hadoop. Besides, not every company needs big data; many do fine with small databases and the traditional relational databases are perfect for that.

If you look at the big data mind map shown in figure 6.1, you'll see four types of NoSQL databases.

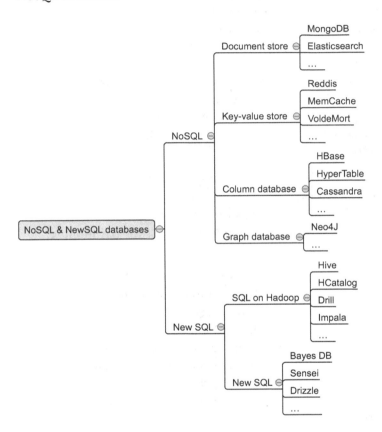

Figure 6.1 NoSQL and NewSQL databases

These four types are document store, key-value store, graph database, and column database. The mind map also includes the NewSQL partitioned relational databases. In the future this big split between NoSQL and NewSQL will become obsolete because every database type will have its own focus, while combining elements from both NoSQL and NewSQL databases. The lines are slowly blurring as RDBMS types get NoSQL features such as the column-oriented indexing seen in columnar databases. But for now it's a good way to show that the old relational databases have moved past their single-node setup, while other database types are emerging under the NoSQL denominator.

Let's look at what NoSQL brings to the table.

6.1   *Introduction to NoSQL*

As you've read, the goal of NoSQL databases isn't only to offer a way to partition databases successfully over multiple nodes, but also to present fundamentally different ways to model the data at hand to fit its structure to its use case and not to how a relational database requires it to be modeled.

To help you understand NoSQL, we're going to start by looking at the core ACID principles of single-server relational databases and show how NoSQL databases rewrite them into BASE principles so they'll work far better in a distributed fashion. We'll also look at the CAP theorem, which describes the main problem with distributing databases across multiple nodes and how ACID and BASE databases approach it.

6.1.1   *ACID: the core principle of relational databases*

The main aspects of a traditional relational database can be summarized by the concept ACID:

- *Atomicity*—The "all or nothing" principle. If a record is put into a database, it's put in completely or not at all. If, for instance, a power failure occurs in the middle of a database write action, you wouldn't end up with half a record; it wouldn't be there at all.
- *Consistency*—This important principle maintains the integrity of the data. No entry that makes it into the database will ever be in conflict with predefined rules, such as lacking a required field or a field being numeric instead of text.
- *Isolation*—When something is changed in the database, nothing can happen on this exact same data at exactly the same moment. Instead, the actions happen in serial with other changes. Isolation is a scale going from low isolation to high isolation. On this scale, traditional databases are on the "high isolation" end. An example of low isolation would be Google Docs: Multiple people can write to a document at the exact same time and see each other's changes happening instantly. A traditional Word document, on the other end of the spectrum, has high isolation; it's locked for editing by the first user to open it. The second person opening the document can view its last saved version but is unable to see unsaved changes or edit the document without first saving it as a copy. So once someone has it opened, the most up-to-date version is completely isolated from anyone but the editor who locked the document.
- *Durability*—If data has entered the database, it should survive permanently. Physical damage to the hard discs will destroy records, but power outages and software crashes should not.

ACID applies to all relational databases and certain NoSQL databases, such as the graph database Neo4j. We'll further discuss graph databases later in this chapter and in chapter 7. For most other NoSQL databases another principle applies: BASE. To understand BASE and why it applies to most NoSQL databases, we need to look at the CAP Theorem.

6.1.2   *CAP Theorem: the problem with DBs on many nodes*

Once a database gets spread out over different servers, it's difficult to follow the ACID principle because of the consistency ACID promises; the CAP Theorem points out why this becomes problematic. The CAP Theorem states that a database can be any two of the following things but never all three:

- *Partition tolerant*—The database can handle a network partition or network failure.
- *Available*—As long as the node you're connecting to is up and running and you can connect to it, the node will respond, even if the connection between the different database nodes is lost.
- *Consistent*—No matter which node you connect to, you'll always see the exact same data.

For a single-node database it's easy to see how it's always available and consistent:

- *Available*—As long as the node is up, it's available. That's all the CAP availability promises.
- *Consistent*—There's no second node, so nothing can be inconsistent.

Things get interesting once the database gets partitioned. Then you need to make a choice between availability and consistency, as shown in figure 6.2.

Let's take the example of an online shop with a server in Europe and a server in the United States, with a single distribution center. A German named Fritz and an American named Freddy are shopping at the same time on that same online shop. They see an item and only one is still in stock: a bronze, octopus-shaped coffee table. Disaster strikes, and communication between the two local servers is temporarily down. If you were the owner of the shop, you'd have two options:

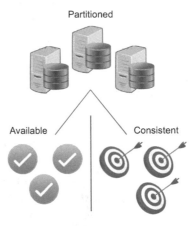

Figure 6.2   CAP Theorem: when partitioning your database, you need to choose between availability and consistency.

- *Availability*—You allow the servers to keep on serving customers, and you sort out everything afterward.
- *Consistency*—You put all sales on hold until communication is reestablished.

In the first case, Fritz and Freddy will both buy the octopus coffee table, because the last-known stock number for both nodes is "one" and both nodes are allowed to sell it, as shown in figure 6.3.

If the coffee table is hard to come by, you'll have to inform either Fritz or Freddy that he won't receive his table on the promised delivery date or, even worse, he will

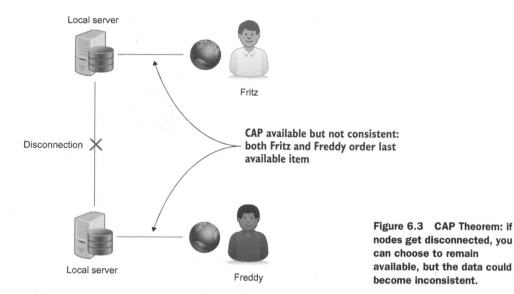

Figure 6.3   CAP Theorem: if nodes get disconnected, you can choose to remain available, but the data could become inconsistent.

never receive it. As a good businessperson, you might compensate one of them with a discount coupon for a later purchase, and everything might be okay after all.

The second option (figure 6.4) involves putting the incoming requests on hold temporarily.

This might be fair to both Fritz and Freddy if after five minutes the web shop is open for business again, but then you might lose both sales and probably many more. Web shops tend to choose availability over consistency, but it's not the optimal choice

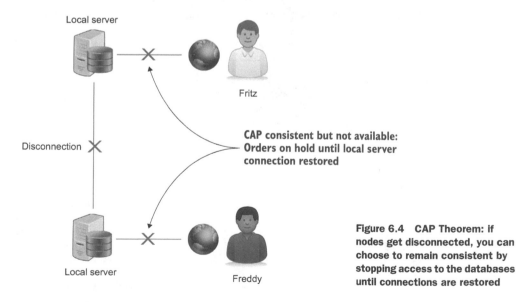

Figure 6.4   CAP Theorem: if nodes get disconnected, you can choose to remain consistent by stopping access to the databases until connections are restored

in all cases. Take a popular festival such as Tomorrowland. Festivals tend to have a maximum allowed capacity for safety reasons. If you sell more tickets than you're allowed because your servers kept on selling during a node communication failure, you could sell double the number allowed by the time communications are reestablished. In such a case it might be wiser to go for consistency and turn off the nodes temporarily. A festival such as Tomorrowland is sold out in the first couple of hours anyway, so a little downtime won't hurt as much as having to withdraw thousands of entry tickets.

6.1.3 *The BASE principles of NoSQL databases*

RDBMS follows the ACID principles; NoSQL databases that don't follow ACID, such as the document stores and key-value stores, follow BASE. BASE is a set of much softer database promises:

- *Basically available*—Availability is guaranteed in the CAP sense. Taking the web shop example, if a node is up and running, you can keep on shopping. Depending on how things are set up, nodes can take over from other nodes. Elasticsearch, for example, is a NoSQL document–type search engine that divides and replicates its data in such a way that node failure doesn't necessarily mean service failure, via the process of *sharding*. Each *shard* can be seen as an individual database server instance, but is also capable of communicating with the other shards to divide the workload as efficiently as possible (figure 6.5). Several shards can be present on a single node. If each shard has a replica on another node, node failure is easily remedied by re-dividing the work among the remaining nodes.

- *Soft state*—The state of a system might change over time. This corresponds to the *eventual consistency principle*: the system might have to change to make the data

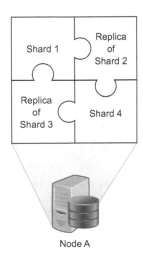

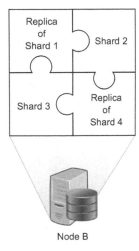

Node A   Node B

Figure 6.5 Sharding: each shard can function as a self-sufficient database, but they also work together as a whole. The example represents two nodes, each containing four shards: two main shards and two replicas. Failure of one node is backed up by the other.

consistent again. In one node the data might say "A" and in the other it might say "B" because it was adapted. Later, at conflict resolution when the network is back online, it's possible the "A" in the first node is replaced by "B." Even though no one did anything to explicitly change "A" into "B," it will take on this value as it becomes consistent with the other node.

- *Eventual consistency*—The database will become consistent over time. In the web shop example, the table is sold twice, which results in data inconsistency. Once the connection between the individual nodes is reestablished, they'll communicate and decide how to resolve it. This conflict can be resolved, for example, on a first-come, first-served basis or by preferring the customer who would incur the lowest transport cost. Databases come with default behavior, but given that there's an actual business decision to make here, this behavior can be overwritten. Even if the connection is up and running, latencies might cause nodes to become inconsistent. Often, products are kept in an online shopping basket, but putting an item in a basket doesn't lock it for other users. If Fritz beats Freddy to the checkout button, there'll be a problem once Freddy goes to check out. This can easily be explained to the customer: he was too late. But what if both press the checkout button at the exact same millisecond and both sales happen?

ACID versus BASE

The BASE principles are somewhat contrived to fit *acid* and *base* from chemistry: an acid is a fluid with a low pH value. A base is the opposite and has a high pH value. We won't go into the chemistry details here, but figure 6.6 shows a mnemonic to those familiar with the chemistry equivalents of acid and base.

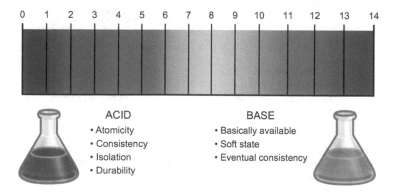

Figure 6.6 ACID versus BASE: traditional relational databases versus most NoSQL databases. The names are derived from the chemistry concept of the pH scale. A pH value below 7 is acidic; higher than 7 is a base. On this scale, your average surface water fluctuates between 6.5 and 8.5.

6.1.4   NoSQL database types

As you saw earlier, there are four big NoSQL types: key-value store, document store, column-oriented database, and graph database. Each type solves a problem that can't be solved with relational databases. Actual implementations are often combinations of these. OrientDB, for example, is a *multi-model database,* combining NoSQL types. OrientDB is a graph database where each node is a document.

Before going into the different NoSQL databases, let's look at relational databases so you have something to compare them to. In data modeling, many approaches are possible. Relational databases generally strive toward *normalization:* making sure every piece of data is stored only once. Normalization marks their structural setup. If, for instance, you want to store data about a person and their hobbies, you can do so with two tables: one about the person and one about their hobbies. As you can see in figure 6.7, an additional table is necessary to link hobbies to persons because of their *many-to-many relationship:* a person can have multiple hobbies and a hobby can have many persons practicing it.

A full-scale relational database can be made up of many entities and linking tables. Now that you have something to compare NoSQL to, let's look at the different types.

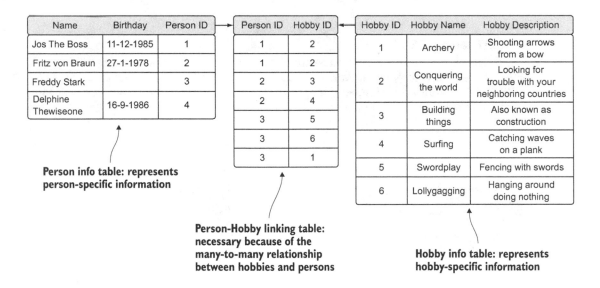

Person info table: represents person-specific information

Person-Hobby linking table: necessary because of the many-to-many relationship between hobbies and persons

Hobby info table: represents hobby-specific information

Figure 6.7   Relational databases strive toward normalization (making sure every piece of data is stored only once). Each table has unique identifiers (primary keys) that are used to model the relationship between the entities (tables), hence the term relational.

COLUMN-ORIENTED DATABASE

Traditional relational databases are row-oriented, with each row having a row id and each field within the row stored together in a table. Let's say, for example's sake, that no extra data about hobbies is stored and you have only a single table to describe people, as shown in figure 6.8. Notice how in this scenario you have slight denormalization because hobbies could be repeated. If the hobby information is a nice extra but not essential to your use case, adding it as a list within the Hobbies column is an acceptable approach. But if the information isn't important enough for a separate table, should it be stored at all?

Row ID	Name	Birthday	Hobbies
1	Jos The Boss	11-12-1985	Archery, conquering the world
2	Fritz von Braun	27-1-1978	Building things, surfing
3	Freddy Stark		Swordplay, lollygagging, archery
4	Delphine Thewiseone	16-9-1986	

Figure 6.8 Row-oriented database layout. Every entity (person) is represented by a single row, spread over multiple columns.

Every time you look up something in a row-oriented database, every row is scanned, regardless of which columns you require. Let's say you only want a list of birthdays in September. The database will scan the table from top to bottom and left to right, as shown in figure 6.9, eventually returning the list of birthdays.

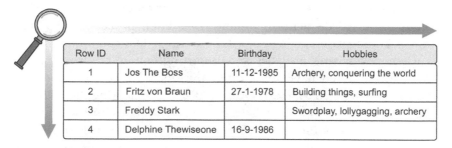

Figure 6.9 Row-oriented lookup: from top to bottom and for every entry, all columns are taken into memory

Indexing the data on certain columns can significantly improve lookup speed, but indexing every column brings extra overhead and the database is still scanning all the columns.

Column databases store each column separately, allowing for quicker scans when only a small number of columns is involved; see figure 6.10.

Name	Row ID
Jos The Boss	1
Fritz von Braun	2
Freddy Stark	3
Delphine Thewiseone	4

Birthday	Row ID
11-12-1985	1
27-1-1978	2
16-9-1986	4

Hobbies	Row ID
Archery	1, 3
Conquering the world	1
Building things	2
Surfing	2
Swordplay	3
Lollygagging	3

Figure 6.10   Column-oriented databases store each column separately with the related row numbers. Every entity (person) is divided over multiple tables.

This layout looks very similar to a row-oriented database with an index on every column. A database *index* is a data structure that allows for quick lookups on data at the cost of storage space and additional writes (index update). An index maps the row number to the data, whereas a column database maps the data to the row numbers; in that way counting becomes quicker, so it's easy to see how many people like archery, for instance. Storing the columns separately also allows for optimized compression because there's only one data type per table.

When should you use a row-oriented database and when should you use a column-oriented database? In a column-oriented database it's easy to add another column because none of the existing columns are affected by it. But adding an entire record requires adapting all tables. This makes the row-oriented database preferable over the column-oriented database for online transaction processing (OLTP), because this implies adding or changing records constantly. The column-oriented database shines when performing analytics and reporting: summing values and counting entries. A row-oriented database is often the operational database of choice for actual transactions (such as sales). Overnight batch jobs bring the column-oriented database up to date, supporting lightning-speed lookups and aggregations using MapReduce algorithms for reports. Examples of column-family stores are Apache HBase, Facebook's Cassandra, Hypertable, and the grandfather of wide-column stores, Google BigTable.

KEY-VALUE STORES

Key-value stores are the least complex of the NoSQL databases. They are, as the name suggests, a collection of key-value pairs, as shown in figure 6.11, and this simplicity makes them the most scalable of the NoSQL database types, capable of storing huge amounts of data.

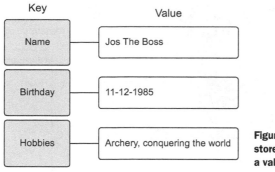

Figure 6.11 Key-value stores store everything as a key and a value.

The value in a key-value store can be anything: a string, a number, but also an entire new set of key-value pairs encapsulated in an object. Figure 6.12 shows a slightly more complex key-value structure. Examples of key-value stores are Redis, Voldemort, Riak, and Amazon's Dynamo.

```
{"internal data":[{"entities":[
    {"customer":[
        {"id":1,"name":"Freddy"},
        {"id":2,"name":"Fritz"}
    ]},
    {"legal entities":[
        {"id":1,"company":"Maiton"}
    ]}]
},{"Products":[
    {"furniture":[
        {"id":1,"name":"Octopus Table","stock":1}
    ]
}]}]}
```

Figure 6.12 Key-value nested structure

DOCUMENT STORES

Document stores are one step up in complexity from key-value stores: a document store does assume a certain document structure that can be specified with a schema. Document stores appear the most natural among the NoSQL database types because they're designed to store everyday documents as is, and they allow for complex querying and calculations on this often already aggregated form of data. The way things are stored in a relational database makes sense from a normalization point of view: everything should be stored only once and connected via foreign keys. Document stores care little about normalization as long as the data is in a structure that makes sense. A relational data model doesn't always fit well with certain business cases. Newspapers or magazines, for example, contain articles. To store these in a relational database, you need to chop them up first: the article text goes in one table, the author and all the information about the author in another, and comments on the article when published on a website go in yet another. As shown in figure 6.13, a newspaper article

Figure 6.13 Document stores save documents as a whole, whereas an RDMS cuts up the article and saves it in several tables. The example was taken from the *Guardian* website.

can also be stored as a single entity; this lowers the cognitive burden of working with the data for those used to seeing articles all the time. Examples of document stores are MongoDB and CouchDB.

GRAPH DATABASES

The last big NoSQL database type is the most complex one, geared toward storing relations between entities in an efficient manner. When the data is highly interconnected, such as for social networks, scientific paper citations, or capital asset clusters, graph databases are the answer. Graph or network data has two main components:

- *Node*—The entities themselves. In a social network this could be people.
- *Edge*—The relationship between two entities. This relationship is represented by a line and has its own properties. An edge can have a direction, for example, if the arrow indicates who is whose boss.

Graphs can become incredibly complex given enough relation and entity types. Figure 6.14 already shows that complexity with only a limited number of entities. Graph databases like Neo4j also claim to uphold ACID, whereas document stores and key-value stores adhere to BASE.

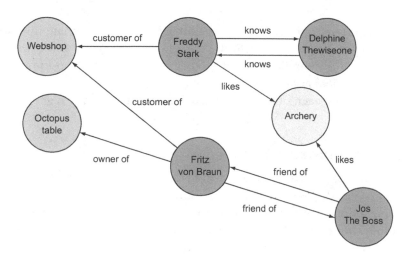

Figure 6.14 Graph data example with four entity types (person, hobby, company, and furniture) and their relations without extra edge or node information

The possibilities are endless, and because the world is becoming increasingly interconnected, graph databases are likely to win terrain over the other types, including the still-dominant relational database. A ranking of the most popular databases and how they're progressing can be found at http://db-engines.com/en/ranking.

257 systems in ranking, March 2015

Rank			DBMS	Database Model	Score		
Mar 2015	Feb 2015	Mar 2014			Mar 2015	Feb 2015	Mar 2014
1.	1.	1.	Oracle	Relational DBMS	1469.09	+29.37	-22.71
2.	2.	2.	MySQL	Relational DBMS	1261.09	-11.36	-29.12
3.	3.	3.	Microsoft SQL Server	Relational DBMS	1164.80	-12.68	-40.48
4.	4.	↑ 5.	MongoDB ➕	Document store	275.01	+7.77	+75.03
5.	5.	↓ 4.	PostgreSQL	Relational DBMS	264.44	+2.10	+29.38
6.	6.	6.	DB2	Relational DBMS	198.85	-3.57	+11.52
7.	7.	7.	Microsoft Access	Relational DBMS	141.69	+1.15	-4.79
8.	8.	↑ 10.	Cassandra ➕	Wide column store	107.31	+0.23	+29.22
9.	9.	↓ 8.	SQLite	Relational DBMS	101.71	+2.14	+8.73
10.	10.	↑ 13.	Redis	Key-value store	97.05	-2.16	+43.59
11.	11.	↓ 9.	SAP Adaptive Server	Relational DBMS	85.37	-0.97	+3.81
12.	12.	12.	Solr	Search engine	81.88	+0.40	+20.74
13.	13.	↓ 11.	Teradata	Relational DBMS	72.78	+3.33	+10.15
14.	14.	↑ 16.	HBase	Wide column store	60.73	+3.59	+25.59
15.	↑ 16.	↑ 19.	Elasticsearch	Search engine	58.92	+6.09	+32.75

Figure 6.15   Top 15 databases ranked by popularity according to DB-Engines.com in March 2015

Figure 6.15 shows that with 9 entries, relational databases still dominate the top 15 at the time this book was written, and with the coming of NewSQL we can't count them out yet. Neo4j, the most popular graph database, can be found at position 23 at the time of writing, with Titan at position 53.

Now that you've seen each of the NoSQL database types, it's time to get your hands dirty with one of them.

6.2   *Case study: What disease is that?*

It has happened to many of us: you have sudden medical symptoms and the first thing you do is Google what disease the symptoms might indicate; then you decide whether it's worth seeing a doctor. A web search engine is okay for this, but a more dedicated database would be better. Databases like this exist and are fairly advanced; they can be almost a virtual version of Dr. House, a brilliant diagnostician in the TV series *House M.D.* But they're built upon well-protected data and not all of it is accessible by the public. Also, although big pharmaceutical companies and advanced hospitals have access to these virtual doctors, many general practitioners are still stuck with only their books. This information and resource asymmetry is not only sad and dangerous, it needn't be there at all. If a simple, disease-specific search engine were used by all general practitioners in the world, many medical mistakes could be avoided.

In this case study, you'll learn how to build such a search engine here, albeit using only a fraction of the medical data that is freely accessible. To tackle the problem, you'll use a modern NoSQL database called Elasticsearch to store the data, and the data science process to work with the data and turn it into a resource that's fast and easy to search. Here's how you'll apply the process:

1 *Setting the research goal.*
2 *Data collection*—You'll get your data from Wikipedia. There are more sources out there, but for demonstration purposes a single one will do.
3 *Data preparation*—The Wikipedia data might not be perfect in its current format. You'll apply a few techniques to change this.
4 *Data exploration*—Your use case is special in that step 4 of the data science process is also the desired end result: you want your data to become easy to explore.
5 *Data modeling*—No real data modeling is applied in this chapter. Document-term matrices that are used for search are often the starting point for advanced topic modeling. We won't go into that here.
6 *Presenting results*—To make data searchable, you'd need a user interface such as a website where people can query and retrieve disease information. In this chapter you won't go so far as to build an actual interface. Your secondary goal: profiling a disease category by its keywords; you'll reach this stage of the data science process because you'll present it as a word cloud, such as the one in figure 6.16.

Figure 6.16   A sample word cloud on non-weighted diabetes keywords

To follow along with the code, you'll need these items:

- A Python session with the elasticsearch-py and Wikipedia libraries installed (`pip install elasticsearch` and `pip install wikipedia`)
- A locally set up Elasticsearch instance; see appendix A for installation instructions
- The IPython library

NOTE   The code for this chapter is available to download from the Manning website for this book at https://manning.com/books/introducing-data-science and is in IPython format.

Elasticsearch: the open source search engine/NoSQL database

To tackle the problem at hand, diagnosing a disease, the NoSQL database you'll use is Elasticsearch. Like MongoDB, Elasticsearch is a document store. But unlike MongoDB, Elasticsearch is a search engine. Whereas MongoDB is great at performing complex calculations and MapReduce jobs, Elasticsearch's main purpose is full-text search. Elasticsearch will do basic calculations on indexed numerical data such as summing, counts, median, mean, standard deviation, and so on, but in essence it remains a search engine.

Elasticsearch is built on top of Apache Lucene, the Apache search engine created in 1999. Lucene is notoriously hard to handle and is more a building block for more user-friendly applications than an end–to–end solution in itself. But Lucene is an enormously powerful search engine, and Apache Solr followed in 2004, opening for public use in 2006. Solr (an open source, enterprise search platform) is built on top of Apache Lucene and is at this moment still the most versatile and popular open source search engine. Solr is a great platform and worth investigating if you get involved in a project requiring a search engine. In 2010 Elasticsearch emerged, quickly gaining in popularity. Although Solr can still be difficult to set up and configure, even for small projects, Elasticsearch couldn't be easier. Solr still has an advantage in the number of possible plugins expanding its core functionality, but Elasticsearch is quickly catching up and today its capabilities are of comparable quality.

6.2.1   Step 1: Setting the research goal

Can you diagnose a disease by the end of this chapter, using nothing but your own home computer and the free software and data out there? Knowing what you want to do and how to do it is the first step in the data science process, as shown in figure 6.17.

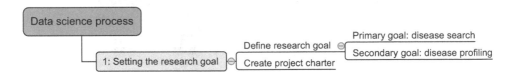

Figure 6.17   Step 1 in the data science process: setting the research goal

- Your primary goal is to set up a disease search engine that would help general practitioners in diagnosing diseases.
- Your secondary goal is to profile a disease: What keywords distinguish it from other diseases?

This secondary goal is useful for educational purposes or as input to more advanced uses such as detecting spreading epidemics by tapping into social media. With your research goal and a plan of action defined, let's move on to the data retrieval step.

6.2.2 Steps 2 and 3: Data retrieval and preparation

Data retrieval and data preparation are two distinct steps in the data science process, and even though this remains true for the case study, we'll explore both in the same section. This way you can avoid setting up local intermedia storage and immediately do data preparation while the data is being retrieved. Let's look at where we are in the data science process (see figure 6.18).

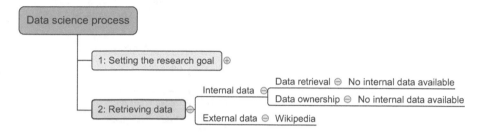

Figure 6.18 Data science process step 2: data retrieval. In this case there's no internal data; all data will be fetched from Wikipedia.

As shown in figure 6.18 you have two possible sources: internal data and external data.

- *Internal data*—You have no disease information lying around. If you currently work for a pharmaceutical company or a hospital, you might be luckier.
- *External data*—All you can use for this case is external data. You have several possibilities, but you'll go with Wikipedia.

When you pull the data from Wikipedia, you'll need to store it in your local Elasticsearch index, but before you do that you'll need to prepare the data. Once data has entered the Elasticsearch index, it can't be altered; all you can do then is query it. Look at the data preparation overview in figure 6.19.

As shown in figure 6.19 there are three distinct categories of data preparation to consider:

- *Data cleansing*—The data you'll pull from Wikipedia can be incomplete or erroneous. Data entry errors and spelling mistakes are possible—even false information isn't excluded. Luckily, you don't need the list of diseases to be exhaustive, and you can handle spelling mistakes at search time; more on that later. Thanks to the Wikipedia Python library, the textual data you'll receive is fairly clean already. If you were to scrape it manually, you'd need to add HTML cleaning, removing all HTML tags. The truth of the matter is full-text search tends to be fairly robust toward common errors such as incorrect values. Even if you dumped in HTML tags on purpose, they'd be unlikely to influence the results; the HTML tags are too different from normal language to interfere.

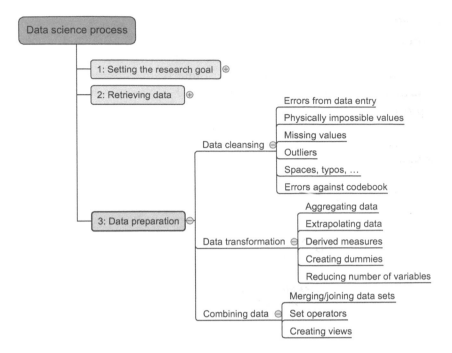

Figure 6.19   Data science process step 3: data preparation

- *Data transformation*—You don't need to transform the data much at this point; you want to search it as is. But you'll make the distinction between page title, disease name, and page body. This distinction is almost mandatory for search result interpretation.

- *Combining data*—All the data is drawn from a single source in this case, so you have no real need to combine data. A possible extension to this exercise would be to get disease data from another source and match the diseases. This is no trivial task because no unique identifier is present and the names are often slightly different.

You can do data cleansing at only two stages: when using the Python program that connects Wikipedia to Elasticsearch and when running the Elasticsearch internal indexing system:

- *Python*—Here you define what data you'll allow to be stored by your document store, but you won't clean the data or transform the data at this stage, because Elasticsearch is better at it for less effort.

- *Elasticsearch*—Elasticsearch will handle the data manipulation (creating the index) under the hood. You can still influence this process, and you'll do so more explicitly later in this chapter.

Now that you have an overview of the steps to come, let's get to work. If you followed the instructions in the appendix, you should now have a local instance of Elasticsearch up and running. First comes data retrieval: you need information on the different diseases. You have several ways to get that kind of data. You could ask companies for their data or get data from Freebase or other open and free data sources. Acquiring your data can be a challenge, but for this example you'll be pulling it from Wikipedia. This is a bit ironic because searches on the Wikipedia website itself are handled by Elasticsearch. Wikipedia used to have its own system build on top of Apache Lucene, but it became unmaintainable, and as of January 2014 Wikipedia began using Elasticsearch instead.

Wikipedia has a Lists of diseases page, as shown in figure 6.20. From here you can borrow the data from the alphabetical lists.

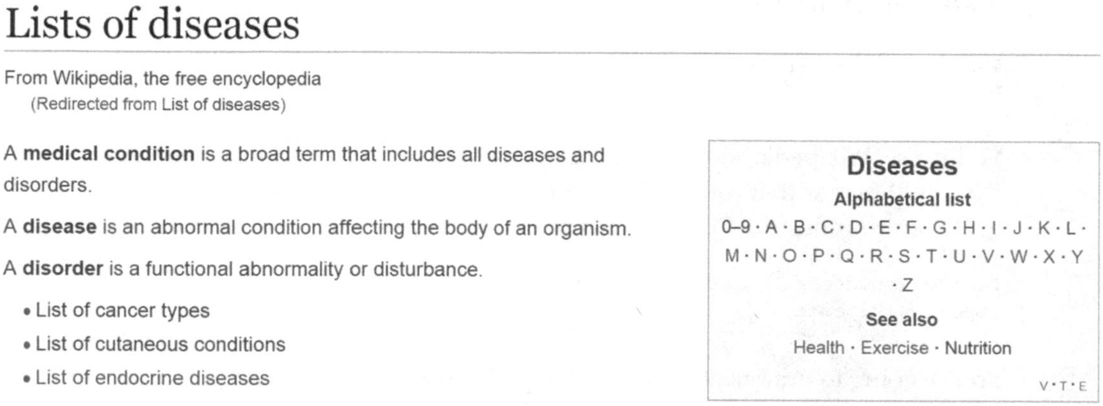

Lists of diseases

From Wikipedia, the free encyclopedia
(Redirected from List of diseases)

A **medical condition** is a broad term that includes all diseases and disorders.

A **disease** is an abnormal condition affecting the body of an organism.

A **disorder** is a functional abnormality or disturbance.

- List of cancer types
- List of cutaneous conditions
- List of endocrine diseases

Diseases

Alphabetical list

0–9 · A · B · C · D · E · F · G · H · I · J · K · L · M · N · O · P · Q · R · S · T · U · V · W · X · Y · Z

See also

Health · Exercise · Nutrition

V · T · E

Figure 6.20   Wikipedia's Lists of diseases page, the starting point for your data retrieval

You know what data you want; now go grab it. You could download the entire Wikipedia data dump. If you want to, you can download it to http://meta.wikimedia.org/wiki/Data_dump_torrents#enwiki.

Of course, if you were to index the entire Wikipedia, the index would end up requiring about 40 GB of storage. Feel free to use this solution, but for the sake of preserving storage and bandwidth, we'll limit ourselves in this book to pulling only the data we intend to use. Another option is scraping the pages you require. Like Google, you can make a program crawl through the pages and retrieve the entire rendered HTML. This would do the trick, but you'd end up with the actual HTML, so you'd need to clean that up before indexing it. Also, unless you're Google, websites aren't too fond of crawlers scraping their web pages. This creates an unnecessarily high amount of traffic, and if enough people send crawlers, it can bring the HTTP server to its knees, spoiling the fun for everyone. Sending billions of requests at the same time is also one of the ways denial of service (DoA) attacks are performed. If you do need to scrape a website, script in a time gap between each page request. This way, your scraper more closely mimics the behavior of a regular website visitor and you won't blow up their servers.

Luckily, the creators of Wikipedia are smart enough to know that this is exactly what would happen with all this information open to everyone. They've put an API in place from which you can safely draw your information. You can read more about it at http://www.mediawiki.org/wiki/API:Main_page.

You'll draw from the API. And Python wouldn't be Python if it didn't already have a library to do the job. There are several actually, but the easiest one will suffice for your needs: Wikipedia.

Activate your Python virtual environment and install all the libraries you'll need for the rest of the book:

```
pip install wikipedia
pip install Elasticsearch
```

You'll use Wikipedia to tap into Wikipedia. Elasticsearch is the main Elasticsearch Python library; with it you can communicate with your database.

Open your favorite Python interpreter and import the necessary libraries:

```
from elasticsearch import Elasticsearch
import wikipedia
```

You're going to draw data from the Wikipedia API and at the same time index on your local Elasticsearch instance, so first you need to prepare it for data acceptance.

```
client = Elasticsearch()          ◁——  Elasticsearch client
indexName = "medical"                   used to communicate
client.indices.create(index=indexName)  with database
                                  ◁——  Index name
                                  ◁——  Create index
```

The first thing you need is a client. `Elasticsearch()` can be initialized with an address but the default is localhost:9200. `Elasticsearch()` and `Elasticsearch('localhost:9200')` are thus the same thing: your client is connected to your local Elasticsearch node. Then you create an index named `"medical"`. If all goes well, you should see an `"acknowledged:true"` reply, as shown in figure 6.21.

Elasticsearch claims to be schema-less, meaning you can use Elasticsearch without defining a database schema and without telling Elasticsearch what kind of data it

```
In [7]:   client = Elasticsearch() #elasticsearch client used to communicate with the database
          indexName = "medical" #the index name
          #cLient.indices.delete(index=indexName) #delete an index
          client.indices.create(index=indexName) #create an index

Out[7]:   {u'acknowledged': True}
```

Figure 6.21   Creating an Elasticsearch index with Python-Elasticsearch

needs to expect. Although this is true for simple cases, you can't avoid having a
schema in the long run, so let's create one, as shown in the following listing.

Listing 6.1   Adding a mapping to the document type

```
diseaseMapping = {                                    ◁─── Defining a mapping
        'properties': {                                    and attributing it to
            'name': {'type': 'string'},                    the disease doc type.
            'title': {'type': 'string'},
            'fulltext': {'type': 'string'}
        }
    }                                                 The "diseases" doc type is
client.indices.put_mapping(index=indexName,           updated with a mapping. Now we
doc_type='diseases',body=diseaseMapping )   ◁────     define the data it should expect.
```

This way you tell Elasticsearch that your index will have a document type called
"disease", and you supply it with the field type for each of the fields. You have three
fields in a disease document: name, title, and fulltext, all of them of type string. If
you hadn't supplied the mapping, Elasticsearch would have guessed their types by
looking at the first entry it received. If it didn't recognize the field to be boolean,
double, float, long, integer, or date, it would set it to string. In this case, you
didn't need to manually specify the mapping.

Now let's move on to Wikipedia. The first thing you want to do is fetch the List of
diseases page, because this is your entry point for further exploration:

```
dl = wikipedia.page("Lists_of_diseases")
```

You now have your first page, but you're more interested in the listing pages because
they contain links to the diseases. Check out the links:

```
dl.links
```

The List of diseases page comes with more links than you'll use. Figure 6.22 shows the
alphabetical lists starting at the sixteenth link.

```
dl = wikipedia.page("Lists_of_diseases")
dl.links
```

```
In [9]:  dl = wikipedia.page("Lists_of_diseases")
         dl.links

Out[9]:  [u'Airborne disease',
          u'Contagious disease',
          u'Cryptogenic disease',
          u'Disease',
          u'Disseminated disease',
          u'Endocrine disease',
          u'Environmental disease',
          u'Eye disease',
          u'Lifestyle disease',
          u'List of abbreviations for diseases and disorders',
          u'List of autism-related topics',
          u'List of basic exercise topics',
          u'List of cancer types',
          u'List of communication disorders',
          u'List of cutaneous conditions',
          u'List of diseases (0\u20139)',
          u'List of diseases (A)',
          u'List of diseases (B)',
```

Figure 6.22 Links on the Wikipedia page Lists of diseases. It has more links than you'll need.

This page has a considerable array of links, but only the alphabetic lists interest you, so keep only those:

```
diseaseListArray = []
for link in dl.links[15:42]:
    try:
        diseaseListArray.append(wikipedia.page(link))
    except Exception,e:
        print str(e)
```

You've probably noticed that the subset is hardcoded, because you know they're the 16th to 43rd entries in the array. If Wikipedia were to add even a single link before the ones you're interested in, it would throw off the results. A better practice would be to use regular expressions for this task. For exploration purposes, hardcoding the entry numbers is fine, but if regular expressions are second nature to you or you intend to turn this code into a batch job, regular expressions are recommended. You can find more information on them at https://docs.python.org/2/howto/regex.html.

One possibility for a regex version would be the following code snippet.

```
diseaseListArray = []
check = re.compile("List of diseases*")
for link in dl.links:
    if check.match(link):
        try:
            diseaseListArray.append(wikipedia.page(link))
        except Exception,e:
            print str(e)
```

```
In [16]:  diseaseListArray

Out[16]:  [<WikipediaPage 'List of diseases (0-9)'>,
           <WikipediaPage 'List of diseases (A)'>,
           <WikipediaPage 'List of diseases (B)'>,
           <WikipediaPage 'List of diseases (C)'>,
           <WikipediaPage 'List of diseases (D)'>,
           <WikipediaPage 'List of diseases (E)'>,
           <WikipediaPage 'List of diseases (F)'>,
           <WikipediaPage 'List of diseases (G)'>,
           <WikipediaPage 'List of diseases (H)'>,
```

Figure 6.23 First Wikipedia disease list, "list of diseases (0-9)"

Figure 6.23 shows the first entries of what you're after: the diseases themselves.

```
diseaseListArray[0].links
```

It's time to index the diseases. Once they're indexed, both data entry and data preparation are effectively over, as shown in the following listing.

Listing 6.2 Indexing diseases from Wikipedia

```
checkList = [["0","1","2","3","4","5","6","7","8","9"],
["A"],["B"],["C"],["D"],["E"],["F"],["G"],["H"],
["I"],["J"],["K"],["L"],["M"],["N"],["O"],["P"],
["Q"],["R"],["S"],["T"],["U"],["V"],["W"],["X"],["Y"],["Z"]]
docType = 'diseases'
for diseaselistNumber, diseaselist in enumerate(diseaseListArray):
    for disease in diseaselist.links:
        try:
            if disease[0] in checkList[diseaselistNumber]
and disease[0:3] !="List":
                currentPage = wikipedia.page(disease)
                client.index(index=indexName,
doc_type=docType,id = disease, body={"name": disease,
    "title":currentPage.title ,
"fulltext":currentPage.content})
        except Exception,e:
            print str(e)
```

The checklist is an array containing an array of allowed first characters. If a disease doesn't comply, skip it.

Document type you'll index.

Looping through disease lists.

Looping through lists of links for every disease list.

First check if it's a disease, then index it.

Because each of the list pages will have links you don't need, check to see if an entry is a disease. You indicate for each list what character the disease starts with, so you check for this. Additionally you exclude the links starting with "list" because these will pop up once you get to the L list of diseases. The check is rather naïve, but the cost of having a few unwanted entries is rather low because the search algorithms will exclude irrelevant results once you start querying. For each disease you index the disease name and the full text of the page. The name is also used as its index ID; this is useful for several advanced Elasticsearch features but also for quick lookup in the browser. For example, try this URL in your browser: http://localhost:9200/medical/diseases/ 11%20beta%20hydroxylase%20deficiency. The title is indexed separately; in most cases the link name and the page title will be identical and sometimes the title will contain an alternative name for the disease.

With at least a few diseases indexed it's possible to make use of the Elasticsearch URI for simple lookups. Have a look at a full body search for the word *headache* in figure 6.24. You can already do this while indexing; Elasticsearch can update an index and return queries for it at the same time.

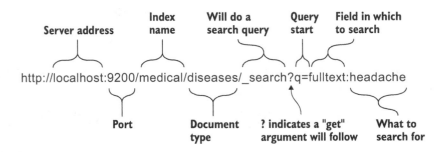

Figure 6.24   The Elasticsearch URL example buildup

If you don't query the index, you can still get a few results without knowing anything about the index. Specifying http://localhost:9200/ medical/diseases/_search will return the first five results. For a more structured view on the data you can ask for the mapping of this document type at http://localhost:9200/medical/ diseases/_mapping?pretty. The pretty get argument shows the returned JSON in a more readable format, as can be seen in figure 6.25. The mapping does appear to be the way you specified it: all fields are type string.

The Elasticsearch URL is certainly useful, yet it won't suffice for your needs. You still have diseases to diagnose, and for this you'll send POST requests to Elasticsearch via your Elasticsearch Python library.

With data retrieval and preparation accomplished, you can move on to exploring your data.

```
← → C ⌂  ⃞ localhost:9200/medical/diseases/_mapping?pretty

{
  "medical" : {
    "mappings" : {
      "diseases" : {
        "properties" : {
          "fulltext" : {
            "type" : "string"
          },
          "name" : {
            "type" : "string"
          },
          "title" : {
            "type" : "string"
          }
        }
      }
    }
  }
}
```

Figure 6.25   Diseases document type mapping via Elasticsearch URL

6.2.3 *Step 4: Data exploration*

> It's not lupus. It's never lupus!
>
> —Dr. House of *House M.D.*

Data exploration is what marks this case study, because the primary goal of the project (disease diagnostics) is a specific way of exploring the data by querying for disease symptoms. Figure 6.26 shows several data exploration techniques, but in this case it's non-graphical: interpreting text search query results.

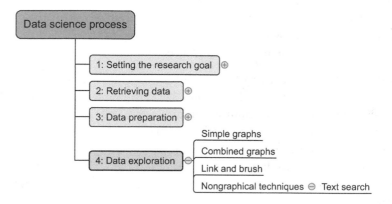

Figure 6.26 Data science process step 4: data exploration

The moment of truth is here: can you find certain diseases by feeding your search engine their symptoms? Let's first make sure you have the basics up and running. Import the Elasticsearch library and define global search settings:

```
from elasticsearch import Elasticsearch
client = Elasticsearch()
indexName = "medical"
docType="diseases"
searchFrom = 0
searchSize= 3
```

You'll return only the first three results; the default is five.

Elasticsearch has an elaborate JSON query language; every search is a POST request to the server and will be answered with a JSON answer. Roughly, the language consists of three big parts: queries, filters, and aggregations. A *query* takes in search keywords and puts them through one or more analyzers before the words are looked up in the index. We'll get deeper into analyzers a bit later in this chapter. A *filter* takes keywords like a query does but doesn't try to analyze what you give it; it filters on the conditions we provide. Filters are thus less complex but many times more efficient because they're also temporarily stored within Elasticsearch in case you use the same filter twice. *Aggregations* can be compared to the SQL group; buckets of words will be created, and for each bucket relevant statistics can be calculated. Each of these three compartments has loads of options and features, making elaborating on the entire language here impossible. Luckily, there's no need to go into the complexity that Elasticsearch queries can represent. We'll use the "Query string query language," a way to query the data that closely resembles the Google search query language. If, for instance, you want a search term to be mandatory, you add a plus (+) sign; if you want to exclude the search term, you use a minus (-) sign. Querying Elasticsearch isn't recommended because it decreases performance; the search engine first needs to translate the query string into its native JSON query language. But for your purposes it will work nicely; also, performance won't be a factor on the several thousand records you have in your index. Now it's time to query your disease data.

PROJECT PRIMARY OBJECTIVE: DIAGNOSING A DISEASE BY ITS SYMPTOMS

If you ever saw the popular television series *House M.D.*, the sentence "It's never lupus" may sound familiar. Lupus is a type of autoimmune disease, where the body's immune system attacks healthy parts of the body. Let's see what symptoms your search engine would need to determine that you're looking for lupus.

Start off with three symptoms: fatigue, fever, and joint pain. Your imaginary patient has all three of them (and more), so make them all mandatory by adding a plus sign before each one:

Listing 6.3 "simple query string" Elasticsearch query with three mandatory keywords

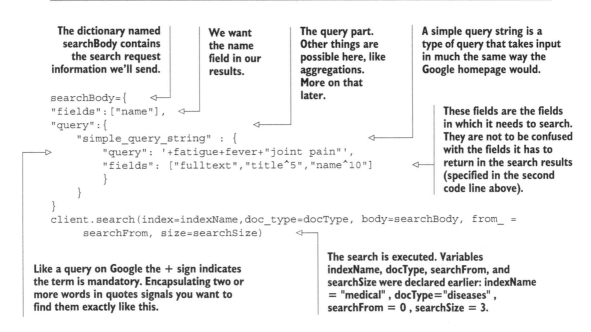

The dictionary named searchBody contains the search request information we'll send.

We want the name field in our results.

The query part. Other things are possible here, like aggregations. More on that later.

A simple query string is a type of query that takes input in much the same way the Google homepage would.

These fields are the fields in which it needs to search. They are not to be confused with the fields it has to return in the search results (specified in the second code line above).

```
searchBody={
"fields":["name"],
"query":{
    "simple_query_string" : {
        "query": '+fatigue+fever+"joint pain"',
        "fields": ["fulltext","title^5","name^10"]
        }
    }
}
client.search(index=indexName,doc_type=docType, body=searchBody, from_ =
    searchFrom, size=searchSize)
```

Like a query on Google the + sign indicates the term is mandatory. Encapsulating two or more words in quotes signals you want to find them exactly like this.

The search is executed. Variables indexName, docType, searchFrom, and searchSize were declared earlier: indexName = "medical" , docType="diseases" , searchFrom = 0 , searchSize = 3.

In searchBody, which has a JSON structure, you specify the fields you'd like to see returned, in this case the name of the disease should suffice. You use the query string syntax to search in all the indexed fields: fulltext, title, and name. By adding ^ you can give each field a weight. If a symptom occurs in the title, it's five times more important than in the open text; if it occurs in the name itself, it's considered ten times as important. Notice how "joint pain" is wrapped in a pair of quotation marks. If you didn't have the "" signs, *joint* and *pain* would have been considered as two separate keywords rather than a single phrase. In Elasticsearch this is called *phrase matching*. Let's look at the results in figure 6.27.

```
{u'_shards': {u'failed': 0, u'successful': 5, u'total': 5},
 u'hits': {u'hits': [{u'_id': u'Macrophagic myofasciitis',
    u'_index': u'medical',
    u'_score': 0.014184786,
    u'_type': u'diseases',
    u'fields': {u'name': [u'Macrophagic myofasciitis']}},
   {u'_id': u'Human granulocytic ehrlichiosis',
    u'_index': u'medical',
    u'_score': 0.0072817733,
    u'_type': u'diseases',
    u'fields': {u'name': [u'Human granulocytic ehrlichiosis']}},
   {u'_id': u'Panniculitis',
    u'_index': u'medical',
    u'_score': 0.0058474476,
    u'_type': u'diseases',
    u'fields': {u'name': [u'Panniculitis']}}],
  u'max_score': 0.014184786,
  u'total': 34},
 u'timed_out': False,
 u'took': 106}
```

Lupus is not in the top 3 diseases returned.

34 diseases found

Figure 6.27　Lupus first search with 34 results

Figure 6.27 shows the top three results returned out of 34 matching diseases. The results are sorted by their matching score, the variable _score. The matching score is no simple thing to explain; it takes into consideration how well the disease matches your query and how many times a keyword was found, the weights you gave, and so on. Currently, lupus doesn't even show up in the top three results. Luckily for you, lupus has another distinct symptom: a rash. The rash doesn't always show up on the person's face, but it does happen and this is where lupus got its name: the face rash makes people vaguely resemble a wolf. Your patient has a rash but not the signature rash on the face, so add "rash" to the symptoms without mentioning the face.

```
"query": '+fatigue+fever+"joint pain"+rash',
```

```
{u'_shards': {u'failed': 0, u'successful': 5, u'total': 5},
 u'hits': {u'hits': [{u'_id': u'Human granulocytic ehrlichiosis',
    u'_index': u'medical',
    u'_score': 0.009902062,
    u'_type': u'diseases',
    u'fields': {u'name': [u'Human granulocytic ehrlichiosis']}},
   {u'_id': u'Lupus erythematosus',
    u'_index': u'medical',
    u'_score': 0.009000875,
    u'_type': u'diseases',
    u'fields': {u'name': [u'Lupus erythematosus']}},
   {u'_id': u'Panniculitis',
    u'_index': u'medical',
    u'_score': 0.007950994,
    u'_type': u'diseases',
    u'fields': {u'name': [u'Panniculitis']}}],
  u'max_score': 0.009902062,
  u'total': 6},
 u'timed_out': False,
 u'took': 15}
```

Figure 6.28 Lupus second search attempt with six results and lupus in the top three

The results of the new search are shown in figure 6.28.

Now the results have been narrowed down to six and lupus is in the top three. At this point, the search engine says *Human Granulocytic Ehrlichiosis* (HGE) is more likely. HGE is a disease spread by ticks, like the infamous Lyme disease. By now a capable doctor would have already figured out which disease plagues your patient, because in determining diseases many factors are at play, more than you can feed into your humble search engine. For instance, the rash occurs only in 10% of HGE and in 50% of lupus patients. Lupus emerges slowly, whereas HGE is set off by a tick bite. Advanced machine-learning databases fed with all this information in a more structured way could make a diagnosis with far greater certainty. Given that you need to make do with the Wikipedia pages, you need another symptom to confirm that it's lupus. The patient experiences chest pain, so add this to the list.

```
"query": '+fatigue+fever+"joint pain"+rash+"chest pain"',
```

The result is shown in figure 6.29.

Seems like it's lupus. It took a while to get to this conclusion, but you got there. Of course, you were limited in the way you presented Elasticsearch with the symptoms. You used only either single terms ("fatigue") or literal phrases ("joint pain"). This worked out for this example, but Elasticsearch is more flexible than this. It can take regular expressions and do a fuzzy search, but that's beyond the scope of this book, although a few examples are included in the downloadable code.

```
{u'_shards': {u'failed': 0, u'successful': 5, u'total': 5},
 u'hits': {u'hits': [{u'_id': u'Lupus erythematosus',
    u'_index': u'medical',
    u'_score': 0.010452312,
    u'_type': u'diseases',
    u'fields': {u'name': [u'Lupus erythematosus']}}],
  u'max_score': 0.010452312,
  u'total': 1},
 u'timed_out': False,
 u'took': 11}
```

Figure 6.29 Lupus third search: with enough symptoms to determine it must be lupus

HANDLING SPELLING MISTAKES: DAMERAU-LEVENSHTEIN

Say someone typed "lupsu" instead of "lupus." Spelling mistakes happen all the time and in all types of human-crafted documents. To deal with this data scientists often use Damerau-Levenshtein. The Damerau-Levenshtein distance between two strings is the number of operations required to turn one string into the other. Four operations are allowed to calculate the distance:

- *Deletion*—Delete a character from the string.
- *Insertion*—Add a character to the string.
- *Substitution*—Substitute one character for another. Without the substitution counted as one operation, changing one character into another would take two operations: one deletion and one insertion.
- *Transposition of two adjacent characters*—Swap two adjacent characters.

This last operation (transposition) is what makes the difference between traditional Levenshtein distance and the Damerau-Levenshtein distance. It's this last operation that makes our dyslexic spelling mistake fall within acceptable limits. Damerau-Levenshtein is forgiving of these transposition mistakes, which makes it great for search engines, but it's also used for other things such as calculating the differences between DNA strings.

Figure 6.30 shows how the transformation from "lupsu" to "lupus" is performed with a single transposition.

Figure 6.30 Adjacent character transposition is one of the operations in Damerau-Levenshtein distance. The other three are insertion, deletion, and substitution.

Lupsu ⟶ Lupsu ⟶ Lupus

With just this you've achieved your first objective: *diagnosing a disease*. But let's not forget about your secondary project objective: *disease profiling*.

PROJECT SECONDARY OBJECTIVE: DISEASE PROFILING

What you want is a list of keywords fitting your selected disease. For this you'll use the significant terms aggregation. The score calculation to determine which words are significant is once again a combination of factors, but it roughly boils down to a comparison of the number of times a term is found in the result set as opposed to all the other documents. This way Elasticsearch profiles your result set by supplying the keywords that distinguish it from the other data. Let's do that on diabetes, a common disease that can take many forms:

Listing 6.4   Significant terms Elasticsearch query for "diabetes"

The dictionary named searchBody contains the search request information we'll send.

We want the name field in our results.

A filtered query has two possible components: a query and a filter. The query performs a search while the filter matches exact values only and is therefore way more efficient but restrictive.

The filter part of the filtered query. A query part isn't mandatory; a filter is sufficient.

The query part.

```
searchBody={
"fields":["name"],
"query":{
    "filtered" : {
        "filter": {
            'term': {'name':'diabetes'}
        }
    }
},
"aggregations" : {
        "DiseaseKeywords" : {
            "significant_terms" : { "field" : "fulltext", "size":30 }
        }
    }
}
client.search(index=indexName,doc_type=docType,
body=searchBody, from_ = searchFrom, size=searchSize)
```

We want to filter the name field and keep only if it contains the term diabetes.

DiseaseKeywords is the name we give to our aggregation.

A significant term aggregation can be compared to keyword detection. The internal algorithm looks for words that are "more important" for the selected set of documents than they are in the overall population of documents.

An aggregation can generally be compared to a group by in SQL. It's mostly used to summarize values of a numeric variable over the distinct values within one or more variables.

You see new code here. You got rid of the query string search and used a filter instead. The filter is encapsulated within the query part because search queries can be combined with filters. It doesn't occur in this example, but when this happens, Elasticsearch will first apply the far more efficient filter before attempting the search. If you know you want to search in a subset of your data, it's always a good idea to add a filter to first create this subset. To demonstrate this, consider the following two snippets of code. They yield the same results but they're not the exact same thing.

A simple query string searching for "diabetes" in the disease name:

```
"query":{
    "simple_query_string" : {
        "query": 'diabetes',
        "fields": ["name"]
        }
    }
```

A term filter filtering in all the diseases with "diabetes" in the name:

```
"query":{
    "filtered" : {
        "filter": {
            'term': {'name':'diabetes'}
        }
    }
}
```

Although it won't show on the small amount of data at your disposal, the filter is way faster than the search. A search query will calculate a search score for each of the diseases and rank them accordingly, whereas a filter simply filters out all those that don't comply. A filter is thus far less complex than an actual search: it's either "yes" or "no" and this is evident in the output. The score is 1 for everything; no distinction is made within the result set. The output consists of two parts now because of the significant terms aggregation. Before you only had hits; now you have hits and aggregations. First, have a look at the hits in figure 6.31.

This should look familiar by now with one notable exception: all results have a score of 1. In addition to being easier to perform, a filter is cached by Elasticsearch for

```
u'hits': {u'hits': [{u'_id': u'Diabetes mellitus',
   u'_index': u'medical',
   u'_score': 1.0,
   u'_type': u'diseases',
   u'fields': {u'name': [u Diabetes mellitus']}},
   {u'_id': u'Diabetes insipidus, nephrogenic type 3',
   u'_index': u'medical',
   u'_score': 1.0,
   u'_type': u'diseases',
   u'fields': {u'name': [u Diabetes insipidus, nephrogenic type 3']}},
   {u'_id': u'Ectodermal dysplasia arthrogryposis diabetes mellitus',
   u'_index': u'medical',
   u'_score': 1.0,
   u'_type': u'diseases',
   u'fields': {u'name': [u'Ectodermal dysplasia arthrogryposis diabetes mellitus']}}],
  u'max_score': 1.0,
  u'total': 27},
 u'timed_out': False,
 u'took': 44}
```

Figure 6.31 Hits output of filtered query with the filter "diabetes" on disease name

awhile. This way, subsequent requests with the same filter are even faster, resulting in a huge performance advantage over search queries.

When should you use filters and when search queries? The rule is simple: use filters whenever possible and use search queries for full-text search when a ranking between the results is required to get the most interesting results at the top.

Now take a look at the significant terms in figure 6.32.

```
{u'_shards': {u'failed': 0, u'successful': 5, u'total': 5},
 u'aggregations': {u'DiseaseKeywords': {u'buckets': [{u'bg_count': 18,
     u'doc_count': 9,
     u'key': u'siphon',
     u'score': 62.84567901234568},
    {u'bg_count': 18,
     u'doc_count': 9,
     u'key': u'diabainein',
     u'score': 62.84567901234568},
    {u'bg_count': 18,
     u'doc_count': 9,
     u'key': u'bainein',
     u'score': 62.84567901234568},
    {u'bg_count': 20,
     u'doc_count': 9,
     u'key': u'passer',
     u'score': 56.52777777777778},
    {u'bg_count': 14,
     u'doc_count': 7,
     u'key': u'ndi',
     u'score': 48.87997256515774},
```

Figure 6.32   Diabetes significant terms aggregation, first five keywords

If you look at the first five keywords in figure 6.32 you'll see that the top four are related to the origin of diabetes. The following Wikipedia paragraph offers help:

> The word diabetes (/ˌdaɪ.ə'biːtiːz/ or /ˌdaɪ.ə'biːtis/) comes from Latin diabētēs, which in turn comes from Ancient Greek διαβήτης (diabētēs) which literally means "a passer through; a siphon" [69]. Ancient Greek physician Aretaeus of Cappadocia (fl. 1st century CE) used that word, with the intended meaning "excessive discharge of urine," as the name for the disease [70, 71, 72]. Ultimately, the word comes from Greek διαβαίνειν (diabainein), meaning "to pass through," [69] which is composed of δια- (dia-), meaning "through" and βαίνειν (bainein), meaning "to go" [70]. The word "diabetes" is first recorded in English, in the form diabete, in a medical text written around 1425.
>
> —Wikipedia page Diabetes_mellitus

This tells you where the word *diabetes* comes from: "a passer through; a siphon" in Greek. It also mentions *diabainein* and *bainein*. You might have known that the most relevant keywords for a disease would be the actual definition and origin. Luckily we asked for 30 keywords, so let's pick a few more interesting ones such as *ndi*. *ndi* is a lowercased version of *NDI*, or "Nephrogenic Diabetes Insipidus," the most common acquired form of diabetes. Lowercase keywords are returned because that's how they're stored in the index when we put it through the standard analyzer when indexing. We didn't specify anything at all while indexing, so the standard analyzer was used by default. Other interesting keywords in the top 30 are *avp*, a gene related to diabetes; *thirst*, a symptom of diabetes; and *Amiloride*, a medication for diabetes. These keywords do seem to profile diabetes, but we're missing multi-term keywords; we stored only individual terms in the index because this was the default behavior. Certain words will never show up on their own because they're not used that often but are still significant when used in combination with other terms. Currently we miss out on the relationship between certain terms. Take *avp*, for example; if *avp* were always written in its full form "Nephrogenic Diabetes Insipidus," it wouldn't be picked up. Storing *n-grams* (combinations of *n* number of words) takes up storage space, and using them for queries or aggregations taxes the search server. Deciding where to stop is a balance exercise and depends on your data and use case.

Generally, bigrams (combination of two terms) are useful because meaningful bigrams exist in the natural language, though 10-grams not so much. Bigram key concepts would be useful for disease profiling, but to create those bigram significant term aggregations you'd need them stored as bigrams in your index. As is often the case in data science, you'll need to go back several steps to make a few changes. Let's go back to the data preparation phase.

6.2.4 Step 3 revisited: Data preparation for disease profiling

It shouldn't come as a surprise that you're back to data preparation, as shown in figure 6.33. The data science process is an iterative one, after all. When you indexed your data, you did virtually no data cleansing or data transformations. You can add data cleansing now by, for instance, stop word filtering. *Stop words* are words that are so common that they're often discarded because they can pollute the results. We won't go into stop word filtering (or other data cleansing) here, but feel free to try it yourself.

To index bigrams you need to create your own token filter and text analyzer. A *token filter* is capable of putting transformations on tokens. Your specific token filter

Figure 6.33 Data science process step 3: data preparation. Data cleansing for text can be stop word filtering; data transformation can be lowercasing of characters.

needs to combine tokens to create *n*-grams, also called *shingles.* The default Elastic-search tokenizer is called the standard tokenizer, and it will look for word boundaries, like the space between words, to cut the text into different tokens or terms. Take a look at the new settings for your disease index, as shown in the following listing.

Listing 6.5   Updating Elasticsearch index settings

```
settings={
    "analysis": {
            "filter": {
                "my_shingle_filter": {
                    "type":            "shingle",
                    "min_shingle_size": 2,
                    "max_shingle_size": 2,
                    "output_unigrams":  False
                }
            },
            "analyzer": {
                "my_shingle_analyzer": {
                    "type":            "custom",
                    "tokenizer":       "standard",
                    "filter": [
                        "lowercase",
                        "my_shingle_filter"
                    ]
                }
            }
        }
    }
client.indices.close(index=indexName)
client.indices.put_settings(index=indexName , body = settings)
client.indices.open(index=indexName)
```

> **Before you can change certain settings, the index needs to be closed. After changing the settings, you can reopen the index.**

You create two new elements: the token filter called "my shingle filter" and a new analyzer called "my_shingle_analyzer." Because *n*-grams are so common, Elastic-search comes with a built-in shingle token filter type. All you need to tell it is that you want the bigrams "min_shingle_size" : 2, "max_shingle_size" : 2, as shown in figure 6.34. You could go for trigrams and higher, but for demonstration purposes this will suffice.

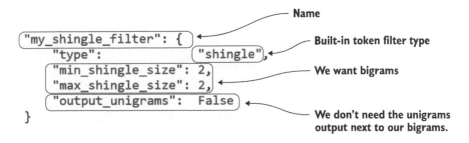

Figure 6.34   A shingle token filter to produce bigrams

The analyzer shown in figure 6.35 is the combination of all the operations required to go from input text to index. It incorporates the shingle filter, but it's much more than this. The tokenizer splits the text into tokens or terms; you can then use a lowercase filter so there's no difference when searching for "Diabetes" versus "diabetes." Finally, you apply your shingle filter, creating your bigrams.

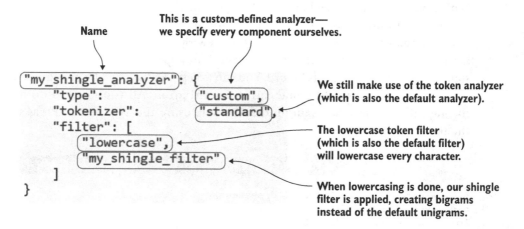

Figure 6.35 **A custom analyzer with standard tokenization and a shingle token filter to produce bigrams**

Notice that you need to close the index before updating the settings. You can then safely reopen the index knowing that your settings have been updated. Not all setting changes require the index to be closed, but this one does. You can find an overview of what settings need the index to be closed at http://www.elastic.co/guide/en/elastic-search/reference/current/indices-update-settings.html.

The index is now ready to use your new analyzer. For this you'll create a new document type, diseases2, with a new mapping, as shown in the following listing.

Listing 6.6 **Create more advanced Elasticsearch doctype mapping**

```
docType = 'diseases2'
diseaseMapping = {
        'properties': {
            'name': {'type': 'string'},
            'title': {'type': 'string'},
            'fulltext': {
                "type": "string",
                "fields": {
                    "shingles": {
                        "type":      "string",
                        "analyzer": "my_shingle_analyzer"
                    }
```

The new disease mapping differs from the old one by the addition of the fulltext.shingles field that contains your bigrams.

```
                }
            }
        }
    }
client.indices.put_mapping(index=indexName,
doc_type=docType,body=diseaseMapping )
```

Within `fulltext` you now have an extra parameter, `fields`. Here you can specify all the different isotopes of `fulltext`. You have only one; it goes by the name `shingles` and will analyze the `fulltext` with your new `my_shingle_analyzer`. You still have access to your original `fulltext`, and you didn't specify an analyzer for this, so the standard one will be used as before. You can access the new one by giving the property name followed by its field name: `fulltext.shingles`. All you need to do now is go through the previous steps and index the data using the Wikipedia API, as shown in the following listing.

Listing 6.7 Reindexing Wikipedia disease explanations with new doctype mapping

```
dl = wikipedia.page("Lists_of_diseases")
diseaseListArray = []
for link in dl.links[15:42]:
    try:
        diseaseListArray.append(wikipedia.page(link))
    except Exception,e:
        print str(e)
```

```
checkList = [["0","1","2","3","4","5","6","7","8","9"],
["A"],["B"],["C"],["D"],["E"],["F"],["G"],
["H"],["I"],["J"],["K"],["L"],["M"],["N"],
["O"],["P"],["Q"],["R"],["S"],["T"],["U"],
["V"],["W"],["X"],["Y"],["Z"]]
```
> The checklist is an array containing allowed "first characters." If a disease doesn't comply, you skip it.

> Loop through disease lists.

```
for diseaselistNumber, diseaselist in enumerate(diseaseListArray):
    for disease in diseaselist.links: #loop through lists of links for every
      disease list
        try:
            if disease[0] in checkList[diseaselistNumber]
and disease[0:3] !="List":
                currentPage = wikipedia.page(disease)
                client.index(index=indexName,
doc_type=docType,id = disease, body={"name": disease,
"title":currentPage.title ,
"fulltext":currentPage.content})
        except Exception,e:
            print str(e)
```
> First check if it's a disease, then index it.

There's nothing new here, only this time you'll index doc_type `diseases2` instead of `diseases`. When this is complete you can again move forward to step 4, data exploration, and check the results.

6.2.5   *Step 4 revisited: Data exploration for disease profiling*

You've once again arrived at data exploration. You can adapt the aggregations query and use your new field to give you bigram key concepts related to diabetes:

Listing 6.8   Significant terms aggregation on "diabetes" with bigrams

```
searchBody={
"fields":["name"],
"query":{
    "filtered" : {
        "filter": {
            'term': {'name':'diabetes'}
        }
    }
},
"aggregations" : {
        "DiseaseKeywords" : {
            "significant_terms" : { "field" : "fulltext", "size" : 30 }
        },
        "DiseaseBigrams": {
            "significant_terms" : { "field" : "fulltext.shingles",
"size" : 30 }
        }
    }
}
client.search(index=indexName,doc_type=docType,
body=searchBody, from_ = 0, size=3)
```

Your new aggregate, called `DiseaseBigrams`, uses the `fulltext.shingles` field to provide a few new insights into diabetes. These new key terms show up:

- *Excessive discharge*—A diabetes patient needs to urinate frequently.
- *Causes polyuria*—This indicates the same thing: diabetes causes the patient to urinate frequently.
- *Deprivation test*—This is actually a trigram, "water deprivation test", but it recognized *deprivation test* because you have only bigrams. It's a test to determine whether a patient has diabetes.
- *Excessive thirst*—You already found "thirst" with your unigram keyword search, but technically at that point it could have meant "no thirst."

There are other interesting bigrams, unigrams, and probably also trigrams. Taken as a whole, they can be used to analyze a text or a collection of texts before reading them. Notice that you achieved the desired results without getting to the modeling stage. Sometimes there's at least an equal amount of valuable information to be found in data exploration as in data modeling. Now that you've fully achieved your secondary objective, you can move on to step 6 of the data science process: presentation and automation.

The rise of graph databases

Where on one hand we're producing data at mass scale, prompting the likes of Google, Amazon, and Facebook to come up with intelligent ways to deal with this, on the other hand we're faced with data that's becoming more interconnected than ever. Graphs and networks are pervasive in our lives. By presenting several motivating examples, we hope to teach the reader how to recognize a graph problem when it reveals itself. In this chapter we'll look at how to leverage those connections for all they're worth using a graph database, and demonstrate how to use Neo4j, a popular graph database.

7.1   *Introducing connected data and graph databases*

Let's start by familiarizing ourselves with the concept of connected data and its representation as graph data.

- *Connected data*—As the name indicates, connected data is characterized by the fact that the data at hand has a relationship that makes it connected.
- *Graphs*—Often referred to in the same sentence as connected data. Graphs are well suited to represent the connectivity of data in a meaningful way.
- *Graph databases*—Introduced in chapter 6. The reason this subject is meriting particular attention is because, besides the fact that data is increasing in size, it's also becoming more interconnected. Not much effort is needed to come up with well-known examples of connected data.

A prominent example of data that takes a network form is social media data. Social media allows us to share and exchange data in networks, thereby generating a great amount of connected data. We can illustrate this with a simple example. Let's assume we have two people in our data, User1 and User2. Furthermore, we know the first name and the last name of User1 (first name: Paul and last name: Beun) and User2 (first name: Jelme and last name: Ragnar). A natural way of representing this could be by drawing it out on a whiteboard, as shown in figure 7.1.

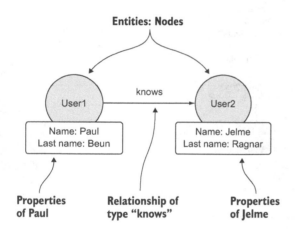

Figure 7.1   A simple connected data example: two entities or nodes (User1, User2), each with properties (first name, last name), connected by a relationship (knows)

The terminology of figure 7.1 is described below:

- *Entities*—We have two entities that represent people (User1 and User2). These entities have the properties "name" and "lastname".
- *Properties*—The properties are defined by key-value pairs. From this graph we can also infer that User1 with the "*name*" property Paul knows User2 with the "*name*" property Jelme.

- *Relationships*—This is the relationship between Paul and Jelme. Note that the relationship has a direction: it's Paul who "*knows*" Jelme and not the other way around. User1 and User2 both represent people and could therefore be grouped.
- *Labels*—In a graph database, one can group nodes by using labels. User1 and User2 could in this case both be labeled as "User".

Connected data often contains many more entities and connections. In figure 7.2 we can see a more extensive graph. Two more entities are included: Country1 with the name Cambodia and Country2 with the name Sweden. Two more relationships exist: "Has_been_in" and "Is_born_in". In the previous graph, only the entities included a property, now the relationships also contain a property. Such graphs are known as property graphs. The relationship connecting the nodes User1 and Country1 is of the type "Has_been_in" and has as a property "Date" which represents a data value. Similarly, User2 is connected to Country2 but through a different type of relationship, which is of the type "Is_born_in". Note that the types of relationships provide us a context of the relationships between nodes. Nodes can have multiple relationships.

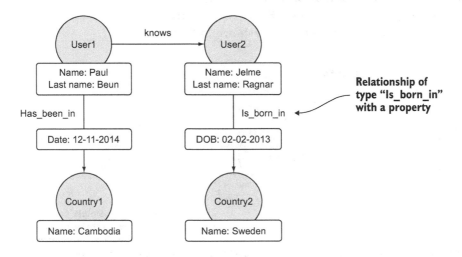

Figure 7.2   A more complicated connected data example where two more entities have been included (Country1 and Country2) and two new relationships ("Has_been_in" and "Is_born_in")

This kind of representation of our data gives us an intuitive way to store connected data. To explore our data we need to traverse through the graph following predefined paths to find the patterns we're searching for. What if one would like to know where Paul has been? Translated into graph database terminology, we'd like to find the pattern "Paul has been in." To answer this, we'd start at the node with the name "Paul" and traverse to Cambodia via the relationship "Has_been_in". Hence a graph traversal, which corresponds to a database query, would be the following:

1 *A starting node*—In this case the node with name property "Paul"
2 *A traversal path*—In this case a path starting at node Paul and going to Cambodia
3 *End node*—Country node with name property "Cambodia"

To better understand how graph databases deal with connected data, it's appropriate to expand a bit more on graphs in general. Graphs are extensively studied in the domains of computer science and mathematics in a field called graph theory. Graph theory is the study of graphs, where graphs represent the mathematical structures used to model pairwise relations between objects, as shown in figure 7.3. What makes them so appealing is that they have a structure that lends itself to visualizing connected data. A graph is defined by vertices (also known as nodes in the graph database world) and edges (also known as relationships). These concepts form the basic fundamentals on which graph data structures are based.

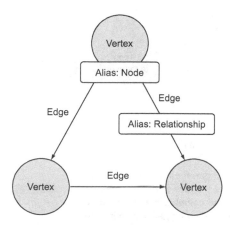

Figure 7.3 At its core a graph consists of nodes (also known as vertices) and edges (that connect the vertices), as known from the mathematical definition of a graph. These collections of objects represent the graph.

Compared to other data structures, a distinctive feature of connected data is its non-linear nature: any entity can be connected to any other via a variety of relationship types and intermediate entities and paths. In graphs, you can make a subdivision between directed and undirected graphs. The edges of a directed graph have—how could it be otherwise—a direction. Although one could argue that every problem could somehow be represented as a graph problem, it's important to understand when it's ideal to do so and when it's not.

7.1.1 Why and when should I use a graph database?

The quest of determining which graph database one should use could be an involved process to undertake. One important aspect in this decision making process is finding the right representation for your data. Since the early 1970s the most common type of database one had to rely on was a relational one. Later, others emerged, such as the hierarchical database (for example, IMS), and the graph database's closest relative: the network database (for example, IDMS). But during the last decades the landscape has become much more diverse, giving end-users more choice depending on their specific needs. Considering the recent development of the data that's becoming available, two characteristics are well suited to be highlighted here. The first one is the size of the data and the other the complexity of the data, as shown in figure 7.4.

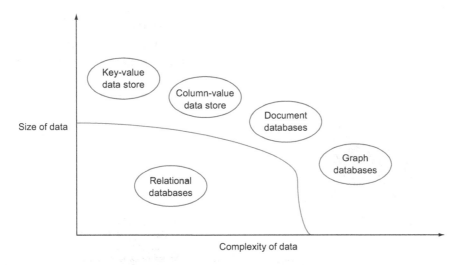

Figure 7.4   This figure illustrates the positioning of graph databases on a two dimensional space where one dimension represents the size of the data one is dealing with, and the other dimension represents the complexity in terms of how connected the data is. When relational databases can no longer cope with the complexity of a data set because of its connectedness, but not its size, graph databases may be your best option.

As figure 7.4 indicates, we'll need to rely on a graph database when the data is complex but still small. Though "small" is a relative thing here, we're still talking hundreds of millions of nodes. Handling complexity is the main asset of a graph database and the ultimate "why" you'd use it. To explain what kind of *complexity* is meant here, first think about how a traditional relational database works.

Contrary to what the name of relational databases indicates, not much is relational about them except that the foreign keys and primary keys are what relate tables. In contrast, relationships in graph databases are first-class citizens. Through this aspect, they lend themselves well to modeling and querying connected data. A relational database would rather strive for minimizing data redundancy. This process is known as database normalization, where a table is decomposed into smaller (less redundant) tables while maintaining all the information intact. In a normalized database one needs to conduct changes of an attribute in only one table. The aim of this process is to isolate data changes in one table. Relational database management systems (RDBMS) are a good choice as a database for data that fits nicely into a tabular format. The relationships in the data can be expressed by joining the tables. Their fit starts to downgrade when the joins become more complicated, especially when they become many-to-many joins. Query time will also increase when your data size starts increasing, and maintaining the database will be more of a challenge. These factors will hamper the performance of your database. Graph databases, on the other hand, inherently store data as nodes and relationships. Although graph databases are classified as a NoSQL type of database, a trend to present them as a category in their own right exists. One seeks the justification for this by noting that the other types of NoSQL databases are aggregation-oriented, while graph databases aren't.

A relational database might, for example, have a table representing "people" and their properties. Any person is related to other people through kinship (and friendship, and so on); each row might represent a person, but connecting them to other rows in the people table would be an immensely difficult job. Do you add a variable that holds the unique identifier of the first child and an extra one to hold the ID of the second child? Where do you stop? Tenth child?

An alternative would be to use an intermediate table for child-parent relationships, but you'll need a separate one for other relationship types like friendship. In this last case you don't get column proliferation but table proliferation: one relationship table for each type of relationship. Even if you somehow succeed in modeling the data in such a way that all family relations are present, you'll need difficult queries to get the answer to simple questions such as "I would like the grandsons of John McBain." First you need to find John McBain's children. Once you find his children, you need to find theirs. By the time you have found all the grandsons, you have hit the "people" table three times:

1 Find McBain and fetch his children.
2 Look up the children with the IDs you got and get the IDs of their children.
3 Find the grandsons of McBain.

Figure 7.5 shows the recursive lookups in a relation database necessary to get from John McBain to his grandsons if everything is in a single table.

Figure 7.6 is another way to model the data: the parent-child relationship is a separate table.

Recursive lookups such as these are inefficient, to say the least.

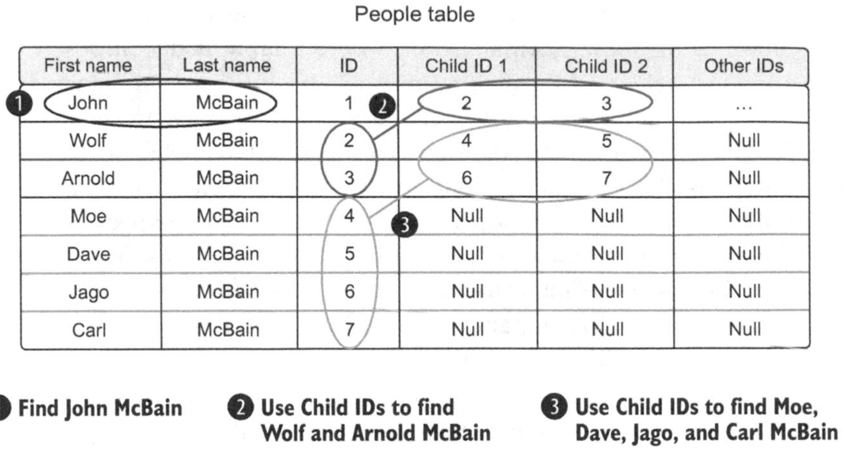

People table

First name	Last name	ID	Child ID 1	Child ID 2	Other IDs
John	McBain	1	2	3	...
Wolf	McBain	2	4	5	Null
Arnold	McBain	3	6	7	Null
Moe	McBain	4	Null	Null	Null
Dave	McBain	5	Null	Null	Null
Jago	McBain	6	Null	Null	Null
Carl	McBain	7	Null	Null	Null

1 Find John McBain   **2** Use Child IDs to find   **3** Use Child IDs to find Moe,
                              Wolf and Arnold McBain   Dave, Jago, and Carl McBain

Figure 7.5   Recursive lookup version 1: all data in one table

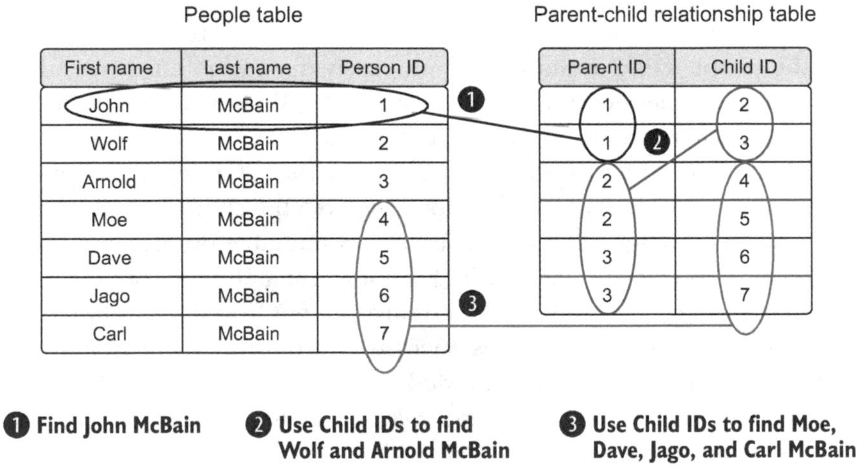

People table   Parent-child relationship table

First name	Last name	Person ID
John	McBain	1
Wolf	McBain	2
Arnold	McBain	3
Moe	McBain	4
Dave	McBain	5
Jago	McBain	6
Carl	McBain	7

Parent ID	Child ID
1	2
1	3
2	4
2	5
3	6
3	7

1 Find John McBain   **2** Use Child IDs to find   **3** Use Child IDs to find Moe,
                              Wolf and Arnold McBain   Dave, Jago, and Carl McBain

Figure 7.6   Recursive lookup version 2: using a parent-child relationship table

Graph databases shine when this type of *complexity* arises. Let's look at the most popular among them.

7.2   Introducing Neo4j: a graph database

Connected data is generally stored in graph databases. These databases are specifically designed to cope with the structure of connected data. The landscape of available graph databases is rather diverse these days. The three most-known ones in order of decreasing popularity are Neo4j, OrientDb, and Titan. To showcase our case study we'll choose the most popular one at the moment of writing (see http://db-engines .com/en/ranking/graph+dbms, September 2015).

Neo4j is a graph database that stores the data in a graph containing nodes and relationships (both are allowed to contain properties). This type of graph database is known as a property graph and is well suited for storing connected data. It has a flexible schema that will give us freedom to change our data structure if needed, providing us the ability to add new data and new relationships if needed. It's an open source project, mature technology, easy to install, user-friendly, and well documented. Neo4j also has a browser-based interface that facilitates the creation of graphs for visualization purposes. To follow along, this would be the right moment to install Neo4j. Neo4j can be downloaded from http://neo4j.com/download/. All necessary steps for a successful installation are summarized in appendix C.

Now let's introduce the four basic structures in Neo4j:

- *Nodes*—Represent entities such as documents, users, recipes, and so on. Certain properties could be assigned to nodes.
- *Relationships*—Exist between the different nodes. They can be accessed either stand-alone or through the nodes they're attached to. Relationships can also contain properties, hence the name property graph model. Every relationship has a name and a direction, which together provide semantic context for the nodes connected by the relationship.
- *Properties*—Both nodes and relationships can have properties. Properties are defined by key-value pairs.
- *Labels*—Can be used to group similar nodes to facilitate faster traversal through graphs.

Before conducting an analysis, a good habit is to design your database carefully so it fits the queries you'd like to run down the road when performing your analysis. Graph databases have the pleasant characteristic that they're whiteboard friendly. If one tries to draw the problem setting on a whiteboard, this drawing will closely resemble the database design for the defined problem. Therefore, such a whiteboard drawing would then be a good starting point to design our database.

Now how to retrieve the data? To explore our data, we need to traverse through the graph following predefined paths to find the patterns we're searching for. The Neo4j browser is an ideal environment to create and play around with your connected data until you get to the right kind of representation for optimal queries, as shown in figure 7.7. The flexible schema of the graph database suits us well here. In this browser you can retrieve your data in rows or as a graph. Neo4j has its own query language to ease the creation and query capabilities of graphs.

Cypher is a highly expressive language that shares enough with SQL to enhance the learning process of the language. In the following section, we'll create our own data using Cypher and insert it into Neo4j. Then we can play around with the data.

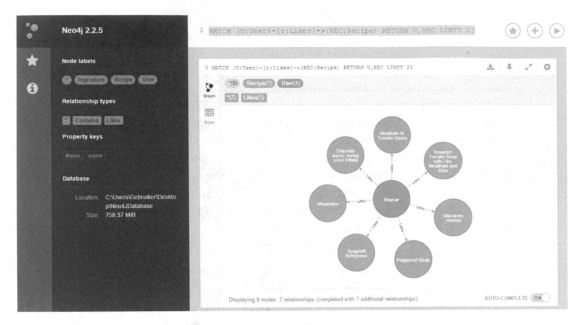

Figure 7.7   Neo4j 2.2.5 interface with resolved query from the chapter case study

7.2.1 *Cypher: a graph query language*

Let's introduce Cypher and its basic syntax for graph operations. The idea of this section is to present enough about Cypher to get us started using the Neo4j browser. At the end of this section you should be able to create your own connected data using Cypher in the Neo4j browser and run basic queries to retrieve the results of the query. For a more extensive introduction to Cypher you can visit http://neo4j.com/docs/stable/cypher-query-lang.html. We'll start by drawing a simple social graph accompanied by a basic query to retrieve a predefined pattern as an example. In the next step we'll draw a more complex graph that will allow us to use more complicated queries in Cypher. This will help us to get acquainted with Cypher and move us down the path to bringing our use case into reality. Moreover, we'll show how to create our own simulated connected data using Cypher.

Figure 7.8 shows a simple social graph of two nodes, connected by a relationship of type "knows". The nodes have both the properties "name" and "lastname".

Now, if we'd like to find out the following pattern, "Who does Paul know?" we'd query this using Cypher. To find a pattern in Cypher, we'll start with a Match clause. In

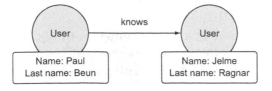

Figure 7.8   An example of a simple social graph with two users and one relationship

this query we'll start searching at the node User with the name property "Paul". Note how the node is enclosed within parentheses, as shown in the code snippet below, and the relationship is enclosed by square brackets. Relationships are named with a colon (:) prefix, and the direction is described using arrows. The placeholder p2 will contain all the User nodes having the relationship of type "knows" as an inbound relationship. With the return clause we can retrieve the results of the query.

```
Match(p1:User { name: 'Paul' } )-[:knows]->(p2:User)
Return p2.name
```

Notice the close relationship of how we have formulated our question verbally and the way the graph database translates this into a traversal. In Neo4j, this impressive expressiveness is made possible by its graph query language, Cypher.

To make the examples more interesting, let's assume that our data is represented by the graph in figure 7.9.

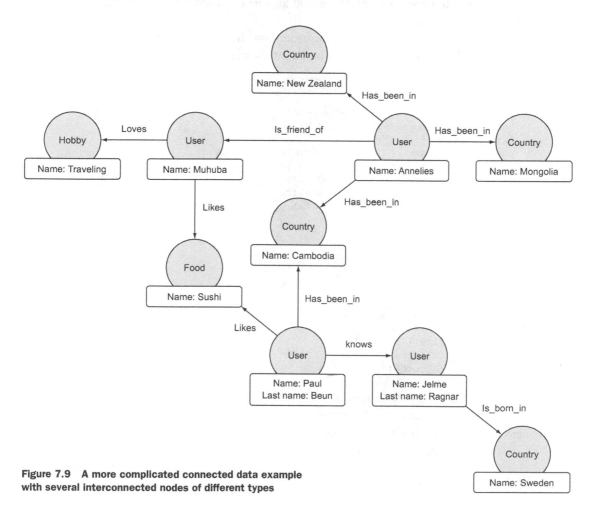

Figure 7.9 A more complicated connected data example with several interconnected nodes of different types

We can insert the connected data in figure 7.9 into Neo4j by using Cypher. We can write Cypher commands directly in the browser-based interface of Neo4j, or alternatively through a Python driver (see http://neo4j.com/developer/python/ for an overview). This is a good way to get a hands-on feeling with connected data and graph databases.

To write an appropriate create statement in Cypher, first we should have a good understanding of which data we'd like to store as nodes and which as relationships, what their properties should be, and whether labels would be useful. The first decision is to decide which data should be regarded as nodes and which as relationships to provide a semantic context for these nodes. In figure 7.9 we've chosen to represent the users and countries they have been in as nodes. Data that provides information about a specific node, for example a name that's associated with a node, can be represented as a property. All data that provides context about two or more nodes will be considered as a relationship. Nodes that share common features, for example Cambodia and Sweden are both countries, will also be grouped through labels. In figure 7.9 this is already done.

In the following listing we demonstrate how the different objects could be encoded in Cypher through one big create statement. *Be aware that Cypher is case sensitive.*

> ### Listing 7.1 Cypher data creation statement

```
CREATE (user1:User {name :'Annelies'}),
 (user2:User {name :'Paul' , LastName: 'Beun'}),
 (user3:User {name :'Muhuba'}),
 (user4:User {name : 'Jelme' , LastName: 'Ragnar'}),
 (country1:Country { name:'Mongolia'}),
 (country2:Country { name:'Cambodia'}),
 (country3:Country { name:'New Zealand'}),
 (country4:Country { name:'Sweden'}),
 (food1:Food { name:'Sushi' }),
 (hobby1:Hobby { name:'Travelling'}),
 (user1)-[:Has_been_in]->(country1),
 (user1)-[: Has_been_in]->(country2),
 (user1)-[: Has_been_in]->(country3),
 (user2)-[: Has_been_in]->(country2),
 (user1)-[: Is_mother_of]->(user4),
 (user2)-[: knows]->(user4),
 (user1)-[: Is_friend_of]->(user3),
 (user2)-[: Likes]->( food1),
 (user3)-[: Likes]->( food1),
 (user4)-[: Is_born_in]->(country4)
```

Running this create statement in one go has the advantage that the success of this execution will ensure us that the graph database has been successfully created. If an error exists, the graph won't be created.

In a real scenario, one should also define indexes and constraints to ensure a fast lookup and not search the entire database. We haven't done this here because our simulated data set is small. However, this can be easily done using Cypher. Consult the

Cypher documentation to find out more about indexes and constraints (http://neo4j.com/docs/stable/cypherdoc-labels-constraints-and-indexes.html). Now that we've created our data, we can query it. The following query will return all nodes and relationships in the database:

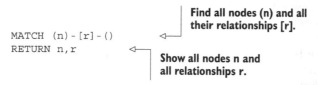

```
MATCH (n)-[r]-()
RETURN n,r
```

Find all nodes (n) and all their relationships [r].

Show all nodes n and all relationships r.

Figure 7.10 shows the database that we've created. We can compare this graph with the graph we've envisioned on our whiteboard. On our whiteboard we grouped nodes of people in a label "User" and nodes of countries in a label "Country". Although the nodes in this figure aren't represented by their labels, the labels are present in our database. Besides that, we also miss a node (Hobby) and a relationship of type "Loves". These can be easily added through a merge statement that will create the node and relationship if they don't exist already:

```
Merge (user3)-[: Loves]->( hobby1)
```

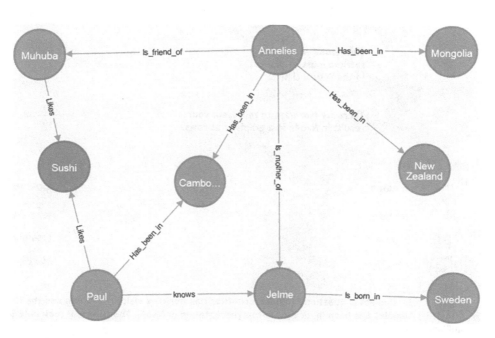

Figure 7.10 The graph drawn in figure 7.9 now has been created in the Neo4j web interface. The nodes aren't represented by their labels but by their names. We can infer from the graph that we're missing the label *Hobby* with the name *Traveling*. The reason for this is because we have forgotten to include this node and its corresponding relationship in the create statement.

We can ask many questions here. For example:

- Question 1: Which countries has Annelies visited? The Cypher code to create the answer (shown in figure 7.11) is

```
Match(u:User{name:'Annelies'}) - [:Has_been_in]-> (c:Country)
Return u.name, c.name
```

- Question 2: Who has been where? The Cypher code (explained in figure 7.12) is

```
Match ()-[r: Has_been_in]->()
Return r LIMIT 25
```

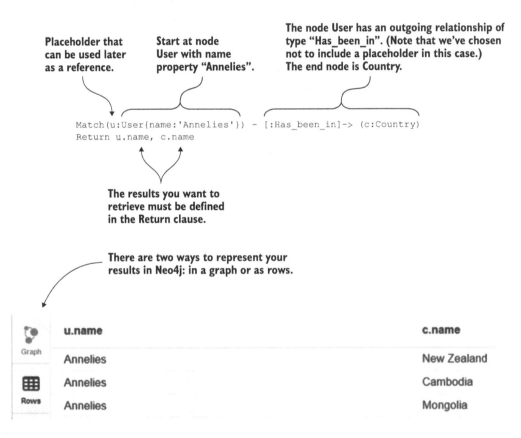

Figure 7.11   Results of question 1: Which countries has Annelies visited? We can see the three countries Annelies has been in, using the row presentation of Neo4j. The traversal took only 97 milliseconds.

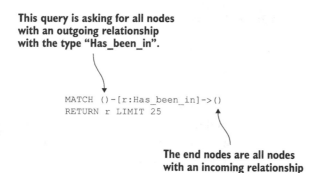

This query is asking for all nodes
with an outgoing relationship
with the type "Has_been_in".

```
MATCH ()-[r:Has_been_in]->()
RETURN r LIMIT 25
```

The end nodes are all nodes
with an incoming relationship
of the type "Has_been_in".

Figure 7.12   Who has been
where? Query buildup explained.

When we run this query we get the answer shown in figure 7.13.

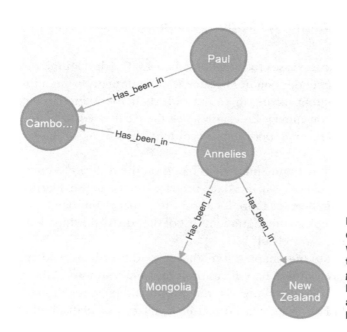

Figure 7.13   Results of
question 2: Who has been
where? The results of our
traversal are now shown in the
graph representation of Neo4j.
Now we can see that Paul, in
addition to Annelies, has also
been to Cambodia.

In question 2 we have chosen not to specify a start node. Therefore, Cypher will go
to all nodes present in the database to find those with an outgoing relationship of
type "Has_been_in". One should avoid not specifying a starting node since, depending on the size of your database, such a query could take a long time to converge.
Playing around with the data to obtain the right graph database also means a lot of
data deletion. Cypher has a delete statement suitable for deleting small amounts of

data. The following query demonstrates how to delete all nodes and relationships in the database:

```
MATCH(n)
Optional MATCH (n)-[r]-()
Delete n,r
```

Now that we're acquainted with connected data and have basic knowledge of how it's managed in a graph database, we can go a step further and look into real, live applications of connected data. A social graph, for example, can be used to find clusters of tightly connected nodes inside the graph communities. People in a cluster who don't know each other can then be introduced to each other. The concept of searching for tightly connected nodes, nodes that have a significant amount of features in common, is a widely used concept. In the next section we'll use this idea, where the aim will be to find clusters inside an ingredient network.

7.3 *Connected data example: a recipe recommendation engine*

One of the most popular use cases for graph databases is the development of recommender engines. Recommender engines became widely adopted through their promise to create relevant content. Living in an era with such abundance in data can be overwhelming to many consumers. Enterprises saw the clear need to be inventive in how to attract customers through personalized content, thereby using the strengths of recommender engines.

In our case study we'll recommend recipes based on the dish preferences of users and a network of ingredients. During data preparation we'll use Elasticsearch to quicken the process and allow for more focus on the actual graph database. Its main purpose here will be to replace the ingredients list of the "dirty" downloaded data with the ingredients from our own "clean" list.

If you skipped ahead to this chapter, it might be good to at least read appendix A on installing Elasticsearch so you have it running on your computer. You can always download the index we'll use from the Manning download page for this chapter and paste it into your local Elasticsearch data directory if you don't feel like bothering with the chapter 6 case study.

You can download the following information from the Manning website for this chapter:

Three .py code files and their .ipynb counterparts

- *Data Preparation Part 1*—Will upload the data to Elasticsearch (alternatively you can paste the downloadable index in your local Elasticsearch data folder)
- *Data Preparation Part 2*—Will move the data from Elasticsearch to Neo4j
- Exploration & Recommender System

Three data files

- *Ingredients (.txt)*—Self-compiled ingredients file
- *Recipes (.json)*—Contains all the ingredients
- *Elasticsearch index (.zip)*—Contains the "gastronomical" Elasticsearch index you can use to skip data preparation part 1

Now that we have everything we need, let's look at the research goal and the steps we need to take to achieve it.

7.3.1   Step 1: Setting the research goal

Let's look at what's to come when we follow the data science process (figure 7.14).

Our primary goal is to set up a recommender engine that would help users of a cooking website find the right recipe. A user gets to like several recipes and we'll base

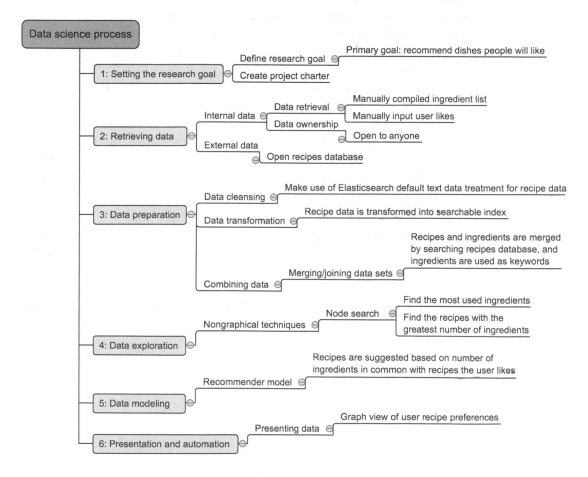

Figure 7.14   Data science process overview applied to connected data recommender model

our dish recommendations on the ingredients' overlap in a recipes network. This is a simple and intuitive approach, yet already yields fairly accurate results. Let's look at the three data elements we require.

7.3.2   *Step 2: Data retrieval*

For this exercise we require three types of data:

- Recipes and their respective ingredients
- A list of distinct ingredients we like to model
- At least one user and his preference for certain dishes

As always, we can divide this into internally available or created data and externally acquired data.

- *Internal data*—We don't have any *user preferences* or ingredients lying around, but these are the smallest part of our data and easily created. A few manually input preferences should be enough to create a recommendation. The user gets more interesting and accurate results the more feedback he gives. We'll input user preferences later in the case study. A *list of ingredients* can be manually compiled and will remain relevant for years to come, so feel free to use the list in the downloadable material for any purpose, commercially or otherwise.
- *External data*—*Recipes* are a different matter. Thousands of ingredients exist, but these can be combined into millions of dishes. We are in luck, however, because a pretty big list is freely available at https://github.com/fictivekin/openrecipes. Many thanks to Fictive Kin for this valuable data set with more than a hundred thousand recipes. Sure there are duplicates in here, but they won't hurt our use case that badly.

We now have two data files at our disposal: a list of 800+ ingredients (ingredients.txt) and more than a hundred thousand recipes in the recipes.json file. A sample of the ingredients list can be seen in the following listing.

Listing 7.2   Ingredients list text file sample

```
Ditalini
Egg Noodles
Farfalle
Fettuccine
Fusilli
Lasagna
Linguine
Macaroni
Orzo
```

The "openrecipes" JSON file contains more than a hundred thousand recipes with multiple properties such as publish date, source location, preparation time, description, and so on. We're only interested in the name and ingredients list. A sample recipe is shown in the following listing.

Listing 7.3 A sample JSON recipe

```
{ "_id" : { "$oid" : "5160756b96cc62079cc2db15" },
    "name" : "Drop Biscuits and Sausage Gravy",
    "ingredients" : "Biscuits\n3 cups All-purpose Flour\n2 Tablespoons Baking
      Powder\n1/2 teaspoon Salt\n1-1/2 stick (3/4 Cup) Cold Butter, Cut Into
      Pieces\n1-1/4 cup Butermilk\n SAUSAGE GRAVY\n1 pound Breakfast Sausage,
      Hot Or Mild\n1/3 cup All-purpose Flour\n4 cups Whole Milk\n1/2 teaspoon
      Seasoned Salt\n2 teaspoons Black Pepper, More To Taste",
    "url" : "http://thepioneerwoman.com/cooking/2013/03/drop-biscuits-and-
      sausage-gravy/",
    "image" : "http://static.thepioneerwoman.com/cooking/files/2013/03/
      bisgrav.jpg",
    "ts" : { "$date" : 1365276011104 },
    "cookTime" : "PT30M",
    "source" : "thepioneerwoman",
    "recipeYield" : "12",
    "datePublished" : "2013-03-11",
    "prepTime" : "PT10M",
    "description" : "Late Saturday afternoon, after Marlboro Man had returned
      home with the soccer-playing girls, and I had returned home with the..."
}
```

Because we're dealing with text data here, the problem is two-fold: first, preparing the textual data as described in the text mining chapter. Then, once the data is thoroughly cleansed, it can be used to produce recipe recommendations based on a network of ingredients. This chapter doesn't focus on the text data preparation because this is described elsewhere, so we'll allow ourselves the luxury of a shortcut during the upcoming data preparation.

7.3.3 Step 3: Data preparation

We now have two data files at our disposal, and we need to combine them into one graph database. The "dirty" recipes data poses a problem that we can address using our clean ingredients list and the use of the search engine and NoSQL database Elasticsearch. We already relied on Elasticsearch in a previous chapter and now it will clean the recipe data for us implicitly when it creates an index. We can then search this data to link each ingredient to every recipe in which it occurs. We could clean the text data using pure Python, as we did in the text mining chapter, but this shows it's good to be aware of the strong points of each NoSQL database; don't pin yourself to a single technology, but use them together to the benefit of the project.

Let's start by entering our recipe data into Elasticsearch. If you don't understand what's happening, please check the case study of chapter 6 again and it should become clear. Make sure to turn on your local Elasticsearch instance and activate a Python environment with the Elasticsearch module installed before running the code snippet in the following listing. It's recommended not to run this code "as is" in Ipython (or Jupyter) because it prints every recipe key to the screen and your browser can handle only so much output. Either turn off the print statements or run in another Python IDE. The code in this snippet can be found in "Data Preparation Part 1.py".

Listing 7.4   Importing recipe data into Elasticsearch

```
from elasticsearch import Elasticsearch          Import
import json                                       modules.

client = Elasticsearch ()            ◁───┐  Elasticsearch client used
indexName = "gastronomical"              │  to communicate with
docType = 'recipes'                      │  database.

                                                        Location of JSON
client.indices.create(index=indexName)   ◁──┐ Create index.   recipe file: change
                                                        this to match your
file_name = 'C:/Users/Gebruiker/Downloads/recipes.json'  ◁──┘ own setup!

recipeMapping = {
      'properties': {
           'name': {'type': 'string'},        Mapping for
           'ingredients': {'type': 'string'}  Elasticsearch
      }                                        "recipe" doctype.
   }

client.indices.put_mapping(index=indexName,doc_type=docType,body=recipeMapping )

with open(file_name, encoding="utf8") as data_file:
    recipeData = json.load(data_file)

for recipe in recipeData:
    print recipe.keys()
    print recipe['_id'].keys()
    client.index(index=indexName,
        doc_type=docType,id = recipe['_id']['$oid'],
        body={"name": recipe['name'], "ingredients":recipe['ingredients']})
```

Load JSON recipe file into memory. **Another way to do this would be:** recipeData = [] with open(file_name) as f: for line in f: recipeData.append(json.loads(line))

Index recipes. Only name and ingredients are important for our use case. In case a timeout problem occurs it's possible to increase the timeout delay by specifying, for example, timeout=30 as an argument.

If everything went well, we now have an Elasticsearch index by the name "gastronomical" populated by thousands of recipes. Notice we allowed for duplicates of the same recipe by not assigning the name of the recipe to be the document key. If, for instance, a recipe is called "lasagna" then this can be a salmon lasagna, beef lasagna, chicken lasagna, or any other type. No single recipe is selected as the prototype lasagna; they are all uploaded to Elasticsearch under the same name: "lasagna". This is a choice, so feel free to decide otherwise. It will have a significant impact, as we'll see later on. The door is now open for a systematic upload to our local graph database. Make sure your local graph database instance is turned on when applying the following code. Our username for this database is the default Neo4j and the password is Neo4ja; make sure to adjust this for your local setup. For this we'll also require a Neo4j-specific Python library called py2neo. If you haven't already, now would be the time to install it to your virtual environment using `pip install py2neo` or `conda install py2neo` when using Anaconda. Again, be advised this code will crash your browser when run directly in Ipython or Jupiter. The code in this listing can be found in "Data Preparation Part 2.py".

> **Listing 7.5   Using the Elasticsearch index to fill the graph database**

```
from elasticsearch import Elasticsearch                          Import
from py2neo import Graph, authenticate, Node, Relationship       modules

client = Elasticsearch ()              Elasticsearch client
indexName = "gastronomical"            used to communicate
docType = 'recipes'                    with database
                                                        Authenticate with
                                                        your own username
                                                        and password
authenticate("localhost:7474", "user", "password")
graph_db = Graph("http://localhost:7474/db/data/")

filename = 'C:/Users/Gebruiker/Downloads/ingredients.txt'
ingredients =[]                                          Ingredients text
with open(filename) as f:                                file gets loaded
    for line in f:                                       into memory
        ingredients.append(line.strip())
                                                Strip because of the /n
print ingredients                               you get otherwise
                                                from reading the .txt

ingredientnumber = 0            Loop through
grandtotal = 0                  ingredients and fetch
for ingredient in ingredients:  Elasticsearch result

    try:
        IngredientNode = graph_db.merge_one("Ingredient","Name",ingredient)
    except:
        continue
                                                        Create node in graph
                                                        database for current
    ingredientnumber +=1                                         ingredient
    searchbody = {
        "size" : 99999999,
        "query": {
            "match_phrase":            Phrase matching used, as
                {                      some ingredients consist
                    "ingredients":{    of multiple words
```

Graph database entity → (label pointing to `graph_db = Graph(...)` line)

```
                    "query":ingredient,
                }
            }
        }
    }
    result = client.search(index=indexName,doc_type=docType,body=searchbody)

    print ingredient
    print ingredientnumber
    print "total: " +  str(result['hits']['total'])

    grandtotal = grandtotal + result['hits']['total']
    print "grand total: " +  str(grandtotal)

    for recipe in result['hits']['hits']:

        try:
            RecipeNode =
    graph_db.merge_one("Recipe","Name",recipe['_source']['name'])
            NodesRelationship = Relationship(RecipeNode, "Contains",
    IngredientNode)
            graph_db.create_unique(NodesRelationship)
            print "added: " + recipe['_source']['name'] + " contains " +
    ingredient

        except:
            continue

    print "*************************************"
```

Loop through recipes found for this particular ingredient

Create relationship between this recipe and ingredient

Create node for each recipe that is not already in graph database

Great, we're now the proud owner of a graph database filled with recipes! It's time for connected data exploration.

7.3.4   Step 4: Data exploration

Now that we have our data where we want it, we can manually explore it using the Neo4j interface at http://localhost:7474/browser/.

Nothing stops you from running your Cypher code in this environment, but Cypher can also be executed via the py2neo library. One interesting question we can pose is which ingredients are occurring the most over all recipes? What are we most likely to get into our digestive system if we randomly selected and ate dishes from this database?

```
from py2neo import Graph, authenticate, Node, Relationship
authenticate("localhost:7474", "user", "password")
graph_db = Graph("http://localhost:7474/db/data/")graph_db.cypher.execute("
    MATCH (REC:Recipe)-[r:Contains]->(ING:Ingredient) WITH ING, count(r) AS num
    RETURN ING.Name as Name, num ORDER BY num DESC LIMIT 10;")
```

The query is created in Cypher and says: for all the recipes and their ingredients, count the number of relations per ingredient and return the ten ingredients with the most relations and their respective counts. The results are shown in figure 7.15.

Most of the top 10 list in figure 7.15 shouldn't come as a surprise. With salt proudly at the top of our list, we shouldn't be shocked to find vascular diseases as the number one killer in most western countries. Another interesting question that comes to mind now is from a different perspective: which recipes require the most ingredients?

	Name	num
1	Salt	53885
2	Oil	42585
3	Sugar	38519
4	Pepper	38118
5	Butter	35610
6	Garlic	29879
7	Flour	28175
8	Olive Oil	25979
9	Onion	24888
10	Cloves	22832

Figure 7.15 Top 10 ingredients that occur in the most recipes

```
from py2neo import Graph, Node, Relationship
graph_db = Graph("http://neo4j:neo4ja@localhost:7474/db/data/")
graph_db.cypher.execute("
    MATCH (REC:Recipe)-[r:Contains]->(ING:Ingredient) WITH REC, count(r) AS num
    RETURN REC.Name as Name, num ORDER BY num DESC LIMIT 10;")
```

The query is almost the same as before, but instead of returning the ingredients, we demand the recipes. The result is figure 7.16.

	Name	num
1	Spaghetti Bolognese	59
2	Chicken Tortilla Soup	56
3	Kedgeree	55
4	Butternut Squash Soup	54
5	Hearty Beef Stew	53
6	Chicken Tikka Masala	52
7	Fish Tacos	52
8	Cooking For Others: 25 Years of Jor, 1 of BGSK	51
9	hibernation fare	50
10	Gazpacho	50

Figure 7.16 Top 10 dishes that can be created with the greatest diversity of ingredients

Now this might be a surprising sight. Spaghetti Bolognese hardly sounds like the type of dish that would require 59 ingredients. Let's take a closer look at the ingredients listed for Spaghetti Bolognese.

```
from py2neo import Graph, Node, Relationship
graph_db = Graph("http://neo4j:neo4ja@localhost:7474/db/data/")
graph_db.cypher.execute("MATCH (REC1:Recipe{Name:'Spaghetti Bolognese'})-
    [r:Contains]->(ING:Ingredient) RETURN REC1.Name, ING.Name;")
```

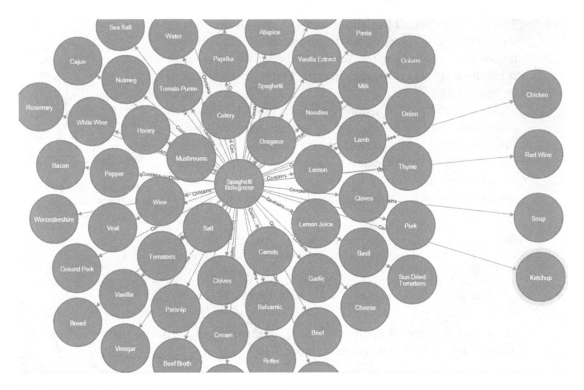

Figure 7.17 Spaghetti Bolognese possible ingredients

The Cypher query merely lists the ingredients linked to Spaghetti Bolognese. Figure 7.17 shows the result in the Neo4j web interface.

Let's remind ourselves of the remark we made when indexing the data in Elasticsearch. A quick Elasticsearch search on Spaghetti Bolognese shows us it occurs multiple times, and all these instances were used to link ingredients to Spaghetti Bolognese as a recipe. We don't have to look at Spaghetti Bolognese as a single recipe but more as a collection of ways people create their own "Spaghetti Bolognese." This makes for an interesting way to look at this data. People can create their version of the dish with ketchup, red wine, and chicken or they might even add soup. With "Spaghetti Bolognese" as a dish being so open to interpretation, no wonder so many people love it.

The Spaghetti Bolognese story was an interesting distraction but not what we came for. It's time to recommend dishes to our gourmand "Ragnar".

7.3.5   *Step 5: Data modeling*

With our knowledge of the data slightly enriched, we get to the goal of this exercise: the recommendations.

For this we introduce a user we call "Ragnar," who likes a couple of dishes. This new information needs to be absorbed by our graph database before we can expect it to suggest new dishes. Therefore, let's now create Ragnar's user node with a few recipe preferences.

> **Listing 7.6 Creating a user node who likes certain recipes in the Neo4j graph database**

```
from py2neo import Graph, Node, Relationship          ← Import modules          Make graph
                                                                                 database
                                                                                 connection
graph_db = Graph("http://neo4j:neo4ja@localhost:7474/db/data/")  ←  object

                                                                     Create
UserRef = graph_db.merge_one("User","Name","Ragnar")                 new user
                                                                     called
                                                                     "Ragnar"
                                                                              Ragnar likes
RecipeRef = graph_db.find_one("Recipe",property_key="Name",         Spaghetti
    property_value="Spaghetti Bolognese")                           Bolognese
NodesRelationship = Relationship(UserRef, "Likes", RecipeRef)   ←
graph_db.create_unique(NodesRelationship) #Commit his like to database

graph_db.create_unique(Relationship(UserRef, "Likes",
    graph_db.find_one("Recipe",property_key="Name",
    property_value="Roasted Tomato Soup with Tiny Meatballs
    and Rice")))
graph_db.create_unique(Relationship(UserRef, "Likes",
    graph_db.find_one("Recipe",property_key="Name",
    property_value="Moussaka")))
graph_db.create_unique(Relationship(UserRef, "Likes",
    graph_db.find_one("Recipe",property_key="Name",
    property_value="Chipolata &amp; spring onion frittata")))
graph_db.create_unique(Relationship(UserRef, "Likes",
    graph_db.find_one("Recipe",property_key="Name",
    property_value="Meatballs In Tomato Sauce")))
graph_db.create_unique(Relationship(UserRef, "Likes",
    graph_db.find_one("Recipe",property_key="Name",
    property_value="Macaroni cheese")))
graph_db.create_unique(Relationship(UserRef, "Likes",
    graph_db.find_one("Recipe",property_key="Name",
    property_value="Peppered Steak")))
```

(margin annotations): Find recipe by the name of Spaghetti Bolognese — Create a like relationship between Ragnar and the spaghetti — Repeat the same process as in the lines above but for several other dishes

In listing 7.6 our food connoisseur Ragnar is added to the database along with his preference for a few dishes. If we select Ragnar in the Neo4j interface, we get figure 7.18. The Cypher query for this is

```
MATCH (U:User)-[r:Likes]->(REC:Recipe) RETURN U,REC LIMIT 25
```

No surprises in figure 7.18: many people like Spaghetti Bolognese, and so does our Scandinavian gastronomist Ragnar.

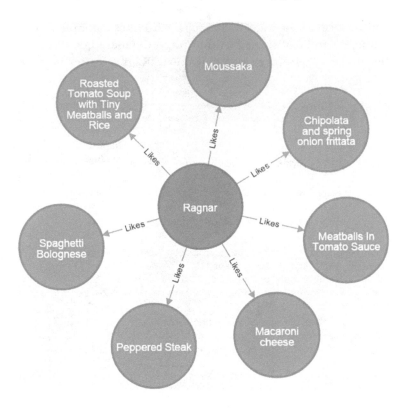

Figure 7.18   The user Ragnar likes several dishes

For the simple recommendation engine we like to build, all that's left for us to do is ask the graph database to give us the nearest dishes in terms of ingredients. Again, this is a basic approach to recommender systems because it doesn't take into account factors such as

- Dislike of an ingredient or a dish.
- The amount of like or dislike. A score out of 10 instead of a binary like or don't like could make a difference.
- The amount of the ingredient that is present in the dish.
- The threshold for a certain ingredient to become apparent in its taste. Certain ingredients, such as spicy pepper, will represent a bigger impact for a smaller dose than other ingredients would.
- Food allergies. While this will be implicitly modeled in the like or dislike of dishes with certain ingredients, a food allergy can be so important that a single mistake can be fatal. Avoidance of allergens should overwrite the entire recommendation system.
- Many more things for you to ponder about.

It might come as a bit of a surprise, but a single Cypher command will suffice.

```
from py2neo import Graph, Node, Relationship
graph_db = Graph("http://neo4j:neo4ja@localhost:7474/db/data/")
graph_db.cypher.execute("
   MATCH (USR1:User{Name:'Ragnar'})-[l1:Likes]->(REC1:Recipe),
         (REC1)-[c1:Contains]->(ING1:Ingredient)
     WITH  ING1,REC1 MATCH (REC2:Recipe)-[c2:Contains]->(ING1:Ingredient)
     WHERE REC1 <> REC2
   RETURN REC2.Name,count(ING1) AS IngCount ORDER BY IngCount DESC LIMIT 20;")
```

First all recipes that **Ragnar** likes are collected. Then their ingredients are used to fetch all the other dishes that share them. The ingredients are then counted for each connected dish and ranked from many common ingredients to few. Only the top 20 dishes are kept; this results in the table of figure 7.19.

```
    | REC2.Name                                                        | IngCount
----+------------------------------------------------------------------+----------
  1 | Spaghetti and Meatballs                                          |      104
  2 | Hearty Beef Stew                                                 |       91
  3 | Cassoulet                                                        |       89
  4 | Lasagne                                                          |       88
  5 | Spaghetti &amp; Meatballs                                        |       86
  6 | Good old lasagne                                                 |       84
  7 | Beef Wellington                                                  |       84
  8 | Braised Short Ribs                                               |       83
  9 | Lasagna                                                          |       83
 10 | Italian Wedding Soup                                             |       82
 11 | French Onion Soup                                                |       82
 12 | Coq au vin                                                       |       82
 13 | Shepherd's pie                                                   |       81
 14 | Great British pork: from head to toe                             |       81
 15 | Three Meat Cannelloni Bake                                       |       81
 16 | Cioppino                                                         |       81
 17 | hibernation fare                                                 |       80
 18 | Spaghetti and Meatballs Recipe with Oven Roasted Tomato Sauce    |       80
 19 | Braised Lamb Shanks                                              |       80
 20 | Lamb and Eggplant Casserole (Moussaka)                           |       80
```

Figure 7.19 Output of the recipe recommendation; top 20 dishes the user may love

From figure 7.19 we can deduce it's time for Ragnar to try Spaghetti and Meatballs, a dish made immortally famous by the Disney animation *Lady and the Tramp.* This does sound like a great recommendation for somebody so fond of dishes containing pasta and meatballs, but as we can see by the ingredient count, many more ingredients back up this suggestion. To give us a small hint of what's behind it, we can show the preferred dishes, the top recommendations, and a few of their overlapping ingredients in a single summary graph image.

7.3.6   Step 6: Presentation

The Neo4j web interface allows us to run the model and retrieve a nice-looking graph that summarizes part of the logic behind the recommendations. It shows how recommended dishes are linked to preferred dishes via the ingredients. This is shown in figure 7.20 and is the final output for our case study.

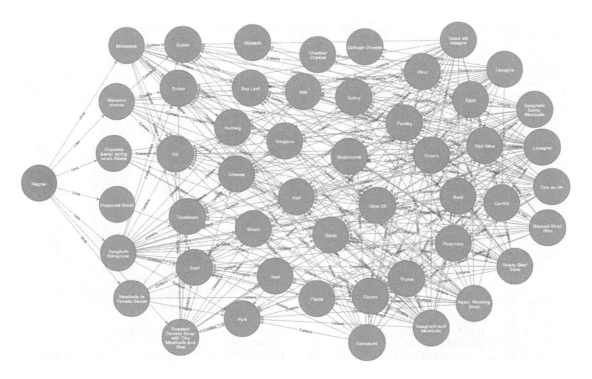

Figure 7.20   Interconnectedness of user-preferred dishes and top 10 recommended dishes via a sub-selection of their overlapping ingredients

With this beautiful graph image we can conclude our chapter in the knowledge that Ragnar has a few tasty dishes to look forward to. Don't forget to try the recommendation system for yourself by inserting your own preferences.

7.4   Summary

In this chapter you learned

- Graph databases are especially useful when encountering data in which relationships between entities are as important as the entities themselves. Compared to the other NoSQL databases, they can handle the biggest complexity but the least data.

- Graph data structures consist of two main components:
  - *Nodes*—These are the entities themselves. In our case study, these are recipes and ingredients.
  - *Edges*—The relationships between entities. Relationships, like nodes, can be of all kinds of types (for example "contains," "likes," "has been to") and can have their own specific properties such as names, weights, or other measures.
- We looked at Neo4j, currently the most popular graph database. For instruction on how to install it, you can consult appendix B. We looked into adding data to Neo4j, querying it using Cypher, and how to access its web interface.
- Cypher is the Neo4j database-specific query language, and we looked at a few examples. We also used it in the case study as part of our dishes recommender system.
- In the chapter's case study we made use of Elasticsearch to clean a huge recipe data dump. We then converted this data to a Neo4j database with recipes and ingredients. The goal of the case study was to recommend dishes to people based on previously shown interest in other dishes. For this we made use of the connectedness of recipes via their ingredients. The py2neo library enabled us to communicate with a Neo4j server from Python.
- It turns out the graph database is not only useful for implementing a recommendation system but also for data exploration. One of the things we found out is the diversity (ingredient-wise) of Spaghetti Bolognese recipes out there.
- We used the Neo4j web interface to create a visual representation of how we get from dish preferences to dish recommendations via the ingredient nodes.

Text mining and text analytics

8

Most of the human recorded information in the world is in the form of written text. We all learn to read and write from infancy so we can express ourselves through writing and learn what others know, think, and feel. We use this skill all the time when reading or writing an email, a blog, text messages, or this book, so it's no wonder written language comes naturally to most of us. Businesses are convinced that much value can be found in the texts that people produce, and rightly so because they contain information on what those people like, dislike, what they know or would like to know, crave and desire, their current health or mood, and so much more. Many of these things can be relevant for companies or researchers, but no single person can read and interpret this tsunami of written material by themself. Once again, we need to turn to computers to do the job for us.

Sadly, however, the natural language doesn't come as "natural" to computers as it does to humans. Deriving meaning and filtering out the unimportant from the important is still something a human is better at than any machine. Luckily, data scientists can apply specific text mining and text analytics techniques to find the relevant information in heaps of text that would otherwise take them centuries to read themselves.

Text mining or *text analytics* is a discipline that combines language science and computer science with statistical and machine learning techniques. Text mining is used for analyzing texts and turning them into a more structured form. Then it takes this structured form and tries to derive insights from it. When analyzing crime from police reports, for example, text mining helps you recognize persons, places, and types of crimes from the reports. Then this new structure is used to gain insight into the evolution of crimes. See figure 8.1.

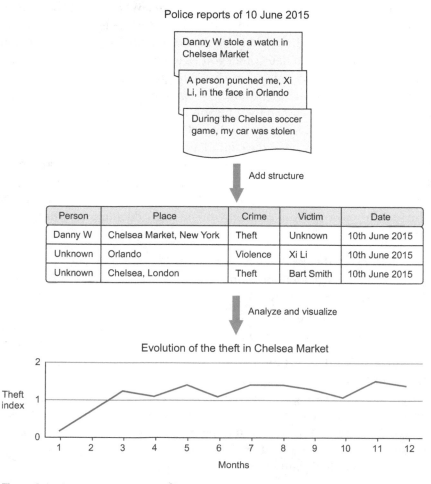

Figure 8.1   In text analytics, (usually) the first challenge is to structure the input text; then it can be thoroughly analyzed.

While language isn't limited to the natural language, the focus of this chapter will be on *Natural Language Processing (NLP)*. Examples of non-natural languages would be machine logs, mathematics, and Morse code. Technically even Esperanto, Klingon, and Dragon language aren't in the field of natural languages because they were invented deliberately instead of evolving over time; they didn't come "natural" to us. These last languages are nevertheless fit for natural communication (speech, writing); they have a grammar and a vocabulary as all natural languages do, and the same text mining techniques could apply to them.

8.1   *Text mining in the real world*

In your day-to-day life you've already come across text mining and natural language applications. Autocomplete and spelling correctors are constantly analyzing the text you type before sending an email or text message. When Facebook autocompletes your status with the name of a friend, it does this with the help of a technique called *named entity recognition,* although this would be only one component of their repertoire. The goal isn't only to detect that you're typing a noun, but also to guess you're referring to a person and recognize who it might be. Another example of named entity recognition is shown in figure 8.2. Google knows Chelsea is a football club but responds differently when asked for a person.

Google uses many types of text mining when presenting you with the results of a query. What pops up in your own mind when someone says "Chelsea"? Chelsea could be many things: a person; a soccer club; a neighborhood in Manhattan, New York or London; a food market; a flower show; and so on. Google knows this and returns different answers to the question "Who is Chelsea?" versus "What is Chelsea?" To provide the most relevant answer, Google must do (among other things) all of the following:

- Preprocess all the documents it collects for named entities
- Perform language identification
- Detect what type of entity you're referring to
- Match a query to a result
- Detect the type of content to return (PDF, adult-sensitive)

This example shows that text mining isn't only about the direct meaning of text itself but also involves meta-attributes such as language and document type.

Google uses text mining for much more than answering queries. Next to shielding its Gmail users from spam, it also divides the emails into different categories such as social, updates, and forums, as shown in figure 8.3.

It's possible to go much further than answering simple questions when you combine text with other logic and mathematics.

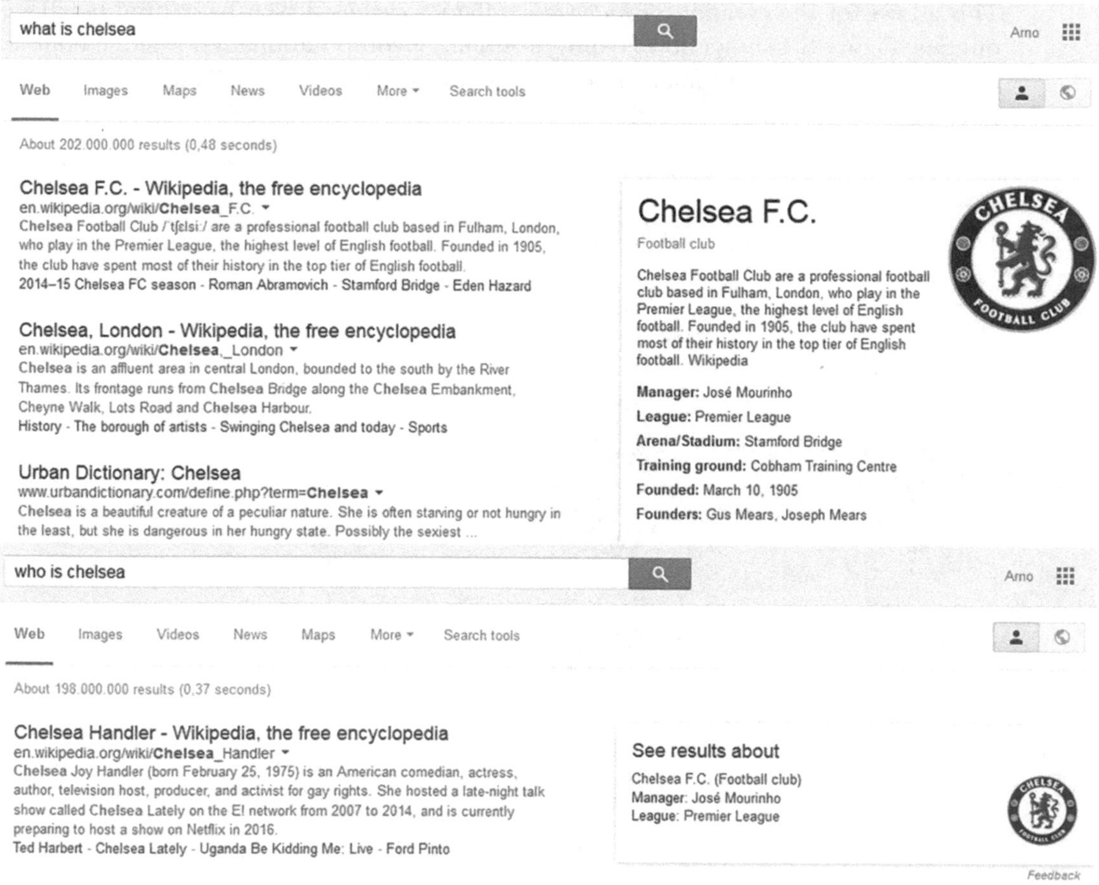

what is chelsea
Arno
Web Images Maps News Videos More Search tools
About 202.000.000 results (0,48 seconds)
Chelsea F.C. - Wikipedia, the free encyclopedia
en.wikipedia.org/wiki/Chelsea_F.C.
Chelsea Football Club /ˈtʃɛlsiː/ are a professional football club based in Fulham, London, who play in the Premier League, the highest level of English football. Founded in 1905, the club have spent most of their history in the top tier of English football.
2014–15 Chelsea FC season - Roman Abramovich - Stamford Bridge - Eden Hazard
Chelsea, London - Wikipedia, the free encyclopedia
en.wikipedia.org/wiki/Chelsea,_London
Chelsea is an affluent area in central London, bounded to the south by the River Thames. Its frontage runs from Chelsea Bridge along the Chelsea Embankment, Cheyne Walk, Lots Road and Chelsea Harbour.
History - The borough of artists - Swinging Chelsea and today - Sports
Urban Dictionary: Chelsea
www.urbandictionary.com/define.php?term=Chelsea
Chelsea is a beautiful creature of a peculiar nature. She is often starving or not hungry in the least, but she is dangerous in her hungry state. Possibly the sexiest ...
Chelsea F.C.
Football club
Chelsea Football Club are a professional football club based in Fulham, London, who play in the Premier League, the highest level of English football. Founded in 1905, the club have spent most of their history in the top tier of English football. Wikipedia
Manager: José Mourinho
League: Premier League
Arena/Stadium: Stamford Bridge
Training ground: Cobham Training Centre
Founded: March 10, 1905
Founders: Gus Mears, Joseph Mears
who is chelsea
Arno
Web Images Videos News Maps More Search tools
About 198.000.000 results (0,37 seconds)
Chelsea Handler - Wikipedia, the free encyclopedia
en.wikipedia.org/wiki/Chelsea_Handler
Chelsea Joy Handler (born February 25, 1975) is an American comedian, actress, author, television host, producer, and activist for gay rights. She hosted a late-night talk show called Chelsea Lately on the E! network from 2007 to 2014, and is currently preparing to host a show on Netflix in 2016.
Ted Harbert - Chelsea Lately - Uganda Be Kidding Me: Live - Ford Pinto
See results about
Chelsea F.C. (Football club)
Manager: José Mourinho
League: Premier League
Feedback

Figure 8.2 The different answers to the queries "Who is Chelsea?" and "What is Chelsea?" imply that Google uses text mining techniques to answer these queries.

Primary
Social
Promotions
Stack Exchange
2 new items in your Stack Exchange inbox - The follo
Stack Exchange
1 new item in your Stack Exchange inbox - The follow

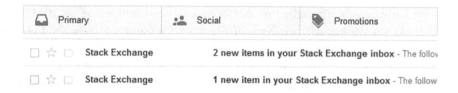

Figure 8.3 Emails can be automatically divided by category based on content and origin.

This allows for the creation of *automatic reasoning engines* driven by natural language queries. Figure 8.4 shows how "Wolfram Alpha," a computational knowledge engine, uses text mining and automatic reasoning to answer the question "Is the USA population bigger than China?"

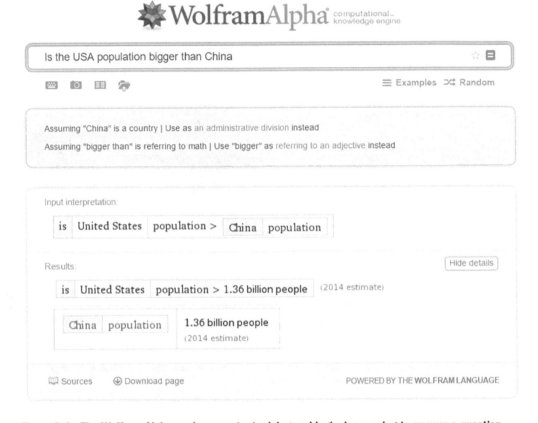

Figure 8.4   The Wolfram Alpha engine uses text mining and logical reasoning to answer a question.

If this isn't impressive enough, the IBM Watson astonished many in 2011 when the machine was set up against two human players in a game of *Jeopardy*. *Jeopardy* is an American quiz show where people receive the answer to a question and points are scored for guessing the correct question for that answer. See figure 8.5.

It's safe to say this round goes to artificial intelligence. IBM Watson is a cognitive engine that can interpret natural language and answer questions based on an extensive knowledge base.

Figure 8.5   IBM Watson wins *Jeopardy* against human players.

Text mining has many applications, including, but not limited to, the following:

- Entity identification
- Plagiarism detection
- Topic identification
- Text clustering
- Translation
- Automatic text summarization
- Fraud detection
- Spam filtering
- Sentiment analysis

Text mining is useful, but is it difficult? Sorry to disappoint: Yes, it is.

When looking at the examples of Wolfram Alpha and IBM Watson, you might have gotten the impression that text mining is easy. Sadly, no. In reality text mining is a complicated task and even many seemingly simple things can't be done satisfactorily. For instance, take the task of guessing the correct address. Figure 8.6 shows how difficult it is to return the exact result with certitude and how Google Maps prompts you for more information when looking for "Springfield." In this case a human wouldn't have done any better without additional context, but this ambiguity is one of the many problems you face in a text mining application.

Figure 8.6 Google Maps asks you for more context due to the ambiguity of the query "Springfield."

Another problem is *spelling mistakes* and *different (correct) spelling* forms of a word. Take the following three references to New York: "NY," "Neww York," and "New York." For a human, it's easy to see they all refer to the city of New York. Because of the way our brain interprets text, understanding text with spelling mistakes comes naturally to us; people may not even notice them. But for a computer these are unrelated strings unless we use algorithms to tell it that they're referring to the same entity. Related problems are synonyms and the use of pronouns. Try assigning the right person to the pronoun "she" in the next sentences: "John gave flowers to Marleen's parents when he met her parents for the first time. She was so happy with this gesture." Easy enough, right? Not for a computer.

We can solve many similar problems with ease, but they often prove hard for a machine. We can train algorithms that work well on a specific problem in a well-defined scope, but more general algorithms that work in all cases are another beast altogether. For instance, we can teach a computer to recognize and retrieve US account numbers from text, but this doesn't generalize well to account numbers from other countries.

Language algorithms are also sensitive to the context the language is used in, even if the language itself remains the same. English models won't work for Arabic and vice versa, but even if we keep to English—an algorithm trained for Twitter data isn't likely to perform well on legal texts. Let's keep this in mind when we move on to the chapter case study: there's no perfect, one-size-fits-all solution in text mining.

8.2 *Text mining techniques*

During our upcoming case study we'll tackle the problem of *text classification*: automatically classifying uncategorized texts into specific categories. To get from raw textual data to our final destination we'll need a few data mining techniques that require background information for us to use them effectively. The first important concept in text mining is the "bag of words."

8.2.1 *Bag of words*

To build our classification model we'll go with the bag of words approach. *Bag of words* is the simplest way of structuring textual data: every document is turned into a word vector. If a certain word is present in the vector it's labeled "True"; the others are labeled "False". Figure 8.7 shows a simplified example of this, in case there are only two documents: one about the television show *Game of Thrones* and one about data science. The two word vectors together form the *document-term matrix*. The document-term matrix holds a column for every term and a row for every document. The values are yours to decide upon. In this chapter we'll use binary: term is present? True or False.

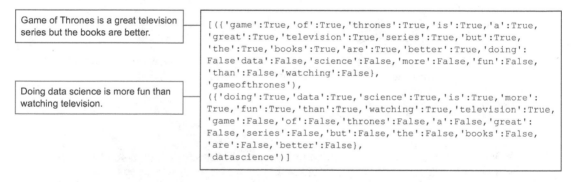

Figure 8.7 A text is transformed into a bag of words by labeling each word (term) with "True" if it is present in the document and "False" if not.

The example from figure 8.7 does give you an idea of the structured data we'll need to start text analysis, but it's severely simplified: not a single word was filtered out and no stemming (we'll go into this later) was applied. A big corpus can have thousands of unique words. If all have to be labeled like this without any filtering, it's easy to see we might end up with a large volume of data. *Binary coded bag of words* as shown in figure 8.7 is but one way to structure the data; other techniques exist.

Term Frequency—Inverse Document Frequency (TF-IDF)

A well-known formula to fill up the document-term matrix is *TF-IDF* or Term Frequency multiplied by Inverse Document Frequency. *Binary bag of words* assigns True or False (term is there or not), while *simple frequencies* count the number of times the term occurred. TF-IDF is a bit more complicated and takes into account how many times a term occurred in the document (TF). TF can be a simple term count, a binary count (True or False), or a logarithmically scaled term count. It depends on what works best for you. In case TF is a term frequency, the formula of TF is the following:

$$TF = f_{t,d}$$

TF is the frequency (f) of the term (t) in the document (d).

But TF-IDF also takes into account all the other documents because of the Inverse Document Frequency. IDF gives an idea of how common the word is in the entire corpus: the higher the document frequency the more common, and more common words are less informative. For example the words "a" or "the" aren't likely to provide specific information on a text. The formula of IDF with logarithmic scaling is the most commonly used form of IDF:

$$IDF = \log(N/|\{d \in D : t \in d\}|)$$

with N being the total number of documents in the corpus, and the $|\{d \in D : t \in d\}|$ being the number of documents (d) in which the term (t) appears.

The TF-IDF score says this about a term: how important is this word to distinguish this document from the others in the corpus? The formula of TF-IDF is thus

$$\frac{1F}{IDF} = f_{t,d}/\log(N/|\{d \in D : t \in d\}|)$$

We won't use TF-IDF, but when setting your next steps in text mining, this should be one of the first things you'll encounter. TF-IDF is also what was used by Elasticsearch behind the scenes in chapter 6. It's a good way to go if you want to use TF-IDF for text analytics; leave the text mining to specialized software such as SOLR or Elasticsearch and take the document/term matrix for text analytics from there.

Before getting to the actual bag of words, many other data manipulation steps take place:

- *Tokenization*—The text is cut into pieces called "tokens" or "terms." These tokens are the most basic unit of information you'll use for your model. The terms are often words but this isn't a necessity. Entire sentences can be used for analysis. We'll use *unigrams*: terms consisting of one word. Often, however, it's useful to include *bigrams* (two words per token) or *trigrams* (three words per token) to capture extra meaning and increase the performance of your models.

This does come at a cost, though, because you're building bigger term-vectors by including bigrams and/or trigrams in the equation.

- *Stop word filtering*—Every language comes with words that have little value in text analytics because they're used so often. NLTK comes with a short list of English stop words we can filter. If the text is tokenized into words, it often makes sense to rid the word vector of these low-information stop words.
- *Lowercasing*—Words with capital letters appear at the beginning of a sentence, others because they're proper nouns or adjectives. We gain no added value making that distinction in our term matrix, so all terms will be set to lowercase.

Another data preparation technique is *stemming*. This one requires more elaboration.

8.2.2 Stemming and lemmatization

Stemming is the process of bringing words back to their root form; this way you end up with less variance in the data. This makes sense if words have similar meanings but are written differently because, for example, one is in its plural form. Stemming attempts to unify by cutting off parts of the word. For example "planes" and "plane" both become "plane."

Another technique, called *lemmatization,* has this same goal but does so in a more grammatically sensitive way. For example, while both stemming and lemmatization would reduce "cars" to "car," lemmatization can also bring back conjugated verbs to their unconjugated forms such as "are" to "be." Which one you use depends on your case, and lemmatization profits heavily from POS Tagging (Part of Speech Tagging). *POS Tagging* is the process of attributing a grammatical label to every part of a sentence. You probably did this manually in school as a language exercise. Take the sentence "*Game of Thrones* is a television series." If we apply POS Tagging on it we get

({"game":"NN"},{"of":"IN"},{"thrones":"NNS"},{"is":"VBZ"},{"a":"DT"},{"television":"NN"}, {"series":"NN"})

NN is a noun, IN is a preposition, NNS is a noun in its plural form, VBZ is a third-person singular verb, and DT is a determiner. Table 8.1 has the full list.

Table 8.1   A list of all POS tags

Tag	Meaning	Tag	Meaning
CC	Coordinating conjunction	CD	Cardinal number
DT	Determiner	EX	Existential
FW	Foreign word	IN	Preposition or subordinating conjunction
JJ	Adjective	JJR	Adjective, comparative
JJS	Adjective, superlative	LS	List item marker
MD	Modal	NN	Noun, singular or mass

Table 8.1   A list of all POS tags *(continued)*

Tag	Meaning	Tag	Meaning
NNS	Noun, plural	NNP	Proper noun, singular
NNPS	Proper noun, plural	PDT	Predeterminer
POS	Possessive ending	PRP	Personal pronoun
PRP$	Possessive pronoun	RB	Adverb
RBR	Adverb, comparative	RBS	Adverb, superlative
RP	Particle	SYM	Symbol
UH	Interjection	VB	Verb, base form
VBD	Verb, past tense	VBG	Verb, gerund or present participle
VBN	Verb, past participle	VBP	Verb, non-3rd person singular present
VBZ	Verb, 3rd person singular present	WDT	Wh-determiner
WP	Wh-pronoun	WP$	Possessive wh-pronoun
WRB	Wh-adverb		

POS Tagging is a use case of sentence-tokenization rather than word-tokenization. After the POS Tagging is complete you can still proceed to word tokenization, but a POS Tagger requires whole sentences. Combining POS Tagging and lemmatization is likely to give cleaner data than using only a stemmer. For the sake of simplicity we'll stick to stemming in the case study, but consider this an opportunity to elaborate on the exercise.

We now know the most important things we'll use to do the data cleansing and manipulation (text mining). For our text analytics, let's add the decision tree classifier to our repertoire.

8.2.3   *Decision tree classifier*

The data analysis part of our case study will be kept simple as well. We'll test a Naïve Bayes classifier and a decision tree classifier. As seen in chapter 3 the Naïve Bayes classifier is called that because it considers each input variable to be independent of all the others, which is naïve, especially in text mining. Take the simple examples of "data science," "data analysis," or "game of thrones." If we cut our data in unigrams we get the following separate variables (if we ignore stemming and such): "data," "science," "analysis," "game," "of," and "thrones." Obviously links will be lost. This can, in turn, be overcome by creating bigrams (data science, data analysis) and trigrams (game of thrones).

The decision tree classifier, however, doesn't consider the variables to be independent of one another and actively creates *interaction variables* and *buckets*. An *interaction variable* is a variable that combines other variables. For instance "data" and "science" might be good predictors in their own right but probably the two of them co-occurring in the same text might have its own value. A bucket is somewhat the opposite. Instead of combining two variables, a variable is split into multiple new ones. This makes sense for numerical variables. Figure 8.8 shows what a decision tree might look like and where you can find interaction and bucketing.

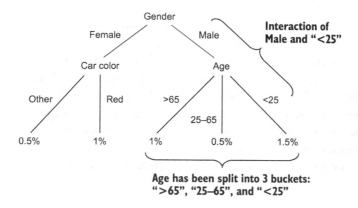

Figure 8.8 **Fictitious decision tree model. A decision tree automatically creates buckets and supposes interactions between input variables.**

Whereas Naïve Bayes supposes independence of all the input variables, a decision tree is built upon the assumption of interdependence. But how does it build this structure? A decision tree has a few possible criteria it can use to split into branches and decide which variables are more important (are closer to the root of the tree) than others. The one we'll use in the NLTK decision tree classifier is "information gain." To understand information gain, we first need to look at entropy. *Entropy* is a measure of unpredictability or chaos. A simple example would be the gender of a baby. When a woman is pregnant, the gender of the fetus can be male or female, but we don't know which one it is. If you were to guess, you have a 50% chance to guess correctly (give or take, because gender distribution isn't 100% uniform). However, during the pregnancy you have the opportunity to do an ultrasound to determine the gender of the fetus. An ultrasound is never 100% conclusive, but the farther along in fetal development, the more accurate it becomes. This accuracy gain, or *information gain*, is there because uncertainty or entropy drops. Let's say an ultrasound at 12 weeks pregnancy has a 90% accuracy in determining the gender of the baby. A 10% uncertainty still exists, but the ultrasound did reduce the uncertainty

Probability of fetus identified as
female—ultrasound at 12 weeks

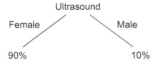

Figure 8.9   Decision tree with one variable: the doctor's conclusion from watching an ultrasound during a pregnancy. What is the probability of the fetus being female?

from 50% to 10%. That's a pretty good discriminator. A decision tree follows this same principle, as shown in figure 8.9.

If another gender test has more predictive power, it could become the root of the tree with the ultrasound test being in the branches, and this can go on until we run out of variables or observations. We can run out of observations, because at every branch split we also split the input data. This is a big weakness of the decision tree, because at the leaf level of the tree robustness breaks down if too few observations are left; the decision trees starts to overfit the data. *Overfitting* allows the model to mistake randomness for real correlations. To counteract this, a decision tree is *pruned*: its meaningless branches are left out of the final model.

Now that we've looked at the most important new techniques, let's dive into the case study.

8.3   *Case study: Classifying Reddit posts*

While text mining has many applications, in this chapter's case study we focus on *document classification*. As pointed out earlier in this chapter, this is exactly what Google does when it arranges your emails in categories or attempts to distinguish spam from regular emails. It's also extensively used by contact centers that process incoming customer questions or complaints: written complaints first pass through a topic detection filter so they can be assigned to the correct people for handling. Document classification is also one of the mandatory features of social media monitoring systems. The monitored tweets, forum or Facebook posts, newspaper articles, and many other internet resources are assigned topic labels. This way they can be reused in reports. *Sentiment analysis* is a specific type of text classification: is the author of a post negative, positive, or neutral on something? That "something" can be recognized with entity recognition.

In this case study we'll draw on posts from Reddit, a website also known as the self-proclaimed "front page of the internet," and attempt to train a model capable of distinguishing whether someone is talking about "data science" or "game of thrones."

The end result can be a presentation of our model or a full-blown interactive application. In chapter 9 we'll focus on application building for the end user, so for now we'll stick to presenting our classification model.

To achieve our goal we'll need all the help and tools we can get, and it happens Python is once again ready to provide them.

8.3.1 *Meet the Natural Language Toolkit*

Python might not be the most execution efficient language on earth, but it has a mature package for text mining and language processing: the *Natural Language Toolkit (NLTK)*. NLTK is a collection of algorithms, functions, and annotated works that will guide you in taking your first steps in text mining and natural language processing. NLTK is also excellently documented on nltk.org. NLTK is, however, not often used for production-grade work, like other libraries such as scikit-learn.

Installing NLTK and its corpora

Install NLTK with your favorite package installer. In case you're using Anaconda, it comes installed with the default Anaconda setup. Otherwise you can go for "pip" or "easy_install". When this is done you still need to install the models and corpora included to have it be fully functional. For this, run the following Python code:

- import nltk
- nltk.download()

Depending on your installation this will give you a pop-up or more command-line options.

Figure 8.10 shows the pop-up box you get when issuing the nltk.download() command.

You can download all the corpora if you like, but for this chapter we'll only make use of "punkt" and "stopwords". This download will be explicitly mentioned in the code that comes with this book.

Figure 8.10 Choose All Packages to fully complete the NLTK installation.

Two IPython notebook files are available for this chapter:

- *Data collection*—Will contain the data collection part of this chapter's case study.
- *Data preparation and analysis*—The stored data is put through data preparation and then subjected to analysis.

All code in the upcoming case study can be found in these two files in the same sequence and can also be run as such. In addition, two interactive graphs are available for download:

- *forceGraph.html*—Represents the top 20 features of our Naïve Bayes model
- *Sunburst.html*—Represents the top four branches of our decision tree model

To open these two HTML pages, an HTTP server is necessary, which you can get using Python and a command window:

- Open a command window (Linux, Windows, whatever you fancy).
- Move to the folder containing the HTML files and their JSON data files: decisionTreeData.json for the sunburst diagram and NaiveBayesData.json for the force graph. It's important the HTML files remain in the same location as their data files or you'll have to change the JavaScript in the HTML file.
- Create a Python HTTP server with the following command: `python -m Simple-HTTPServer 8000`
- Open a browser and go to localhost:8000; here you can select the HTML files, as shown in figure 8.11.

Directory listing for /

- decisionTreeData.json
- forceGraph.html
- NaiveBayesData.json
- sunburst.html

Figure 8.11   Python HTTP server serving this chapter's output

The Python packages we'll use in this chapter:

- *NLTK*—For text mining
- *PRAW*—Allows downloading posts from Reddit
- *SQLite3*—Enables us to store data in the SQLite format
- *Matplotlib*—A plotting library for visualizing data

Make sure to install all the necessary libraries and corpora before moving on. Before we dive into the action, however, let's look at the steps we'll take to get to our goal of creating a topic classification model.

8.3.2 Data science process overview and step 1: The research goal

To solve this text mining exercise, we'll once again make use of the data science process. Figure 8.12 shows the data science process applied to our Reddit classification case.

Not all the elements depicted in figure 8.12 might make sense at this point, and the rest of the chapter is dedicated to working this out in practice as we work toward our research goal: creating a classification model capable of distinguishing posts about "data science" from posts about "Game of Thrones." Without further ado, let's go get our data.

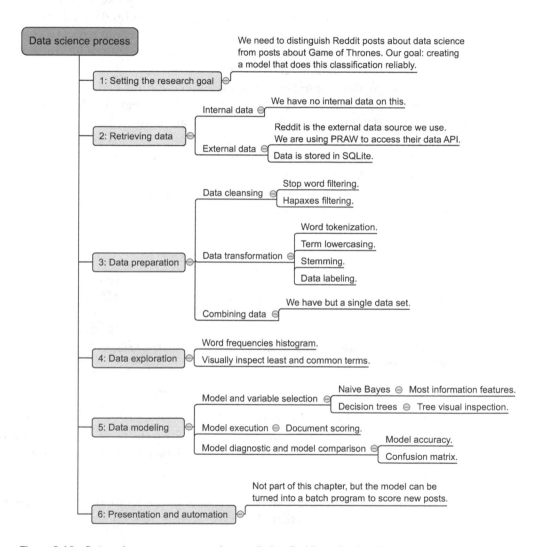

Figure 8.12 Data science process overview applied to Reddit topic classification case study

8.3.3   *Step 2: Data retrieval*

We'll use Reddit data for this case, and for those unfamiliar with Reddit, take the time to familiarize yourself with its concepts at www.reddit.com.

Reddit calls itself "the front page of the internet" because users can post things they find interesting and/or found somewhere on the internet, and only those things deemed interesting by many people are featured as "popular" on its homepage. You could say Reddit gives an overview of the trending things on the internet. Any user can post within a predefined category called a "subreddit." When a post is made, other users get to comment on it and can up-vote it if they like the content or down-vote it if they dislike it. Because a post is always part of a subreddit, we have this metadata at our disposal when we hook up to the Reddit API to get our data. We're effectively fetching labeled data because we'll assume that a post in the subreddit "gameofthrones" has something to do with "gameofthrones."

To get to our data we make use of the official Reddit Python API library called PRAW. Once we get the data we need, we'll store it in a lightweight database-like file called SQLite. SQLite is ideal for storing small amounts of data because it doesn't require any setup to use and will respond to SQL queries like any regular relational database does. Any other data storage medium will do; if you prefer Oracle or Postgres databases, Python has an excellent library to interact with these without the need to write SQL. SQLAlchemy will work for SQLite files as well. Figure 8.13 shows the data retrieval step within the data science process.

Figure 8.13   The data science process data retrieval step for a Reddit topic classification case

Open your favorite Python interpreter; it's time for action, as shown in listing 8.1. First we need to collect our data from the Reddit website. If you haven't already, use `pip install praw` or `conda install praw` (Anaconda) before running the following script.

NOTE   The code for step 2 can also be found in the IPython file "Chapter 8 data collection." It's available in this book's download section.

Listing 8.1 Setting up SQLLite database and Reddit API client

```
import praw          Import PRAW and
import sqlite3       SQLite3 libraries.

conn = sqlite3.connect('reddit.db')      Set up connection to
c = conn.cursor()                        SQLite database.

c.execute('''DROP TABLE IF EXISTS topics''')
c.execute('''DROP TABLE IF EXISTS comments''')
c.execute('''CREATE TABLE topics
            (topicTitle text, topicText text, topicID text,
topicCategory text)''')
c.execute('''CREATE TABLE comments
            (commentText text, commentID text ,
topicTitle text, topicText text, topicID text ,
 topicCategory text)''')
```

Execute SQL statements to create topics and comments table.

```
user_agent = "Introducing Data Science Book"      Create PRAW user agent
r = praw.Reddit(user_agent=user_agent)            so we can use Reddit API.

subreddits = ['datascience','gameofthrones']
```
Our list of subreddits we'll draw into our SQLite database.

```
limit = 1000
```
Maximum number of posts we'll fetch from Reddit per category. Maximum Reddit allows at any single time is also 1,000.

Let's first import the necessary libraries.

Now that we have access to the SQLite3 and PRAW capabilities, we need to prepare our little local database for the data it's about to receive. By defining a connection to a SQLite file we automatically create it if it doesn't already exist. We then define a data cursor that's capable of executing any SQL statement, so we use it to predefine the structure of our database. The database will contain two tables: the topics table contains Reddit topics, which is similar to someone starting a new post on a forum, and the second table contains the comments and is linked to the topic table via the "topicID" column. The two tables have a one (topic table) to many (comment table) relationship. For the case study, we'll limit ourselves to using the topics table, but the data collection will incorporate both because this allows you to experiment with this extra data if you feel like it. To hone your text-mining skills you could perform sentiment analysis on the topic comments and find out what topics receive negative or positive comments. You could then correlate this to the model features we'll produce by the end of this chapter.

We need to create a PRAW client to get access to the data. Every subreddit can be identified by its name, and we're interested in "datascience" and "gameofthrones." The limit represents the maximum number of topics (posts, not comments) we'll draw in from Reddit. A thousand is also the maximum number the API allows us to fetch at any given request, though we could request more later on when people have posted new things. In fact we can run the API request periodically and gather data over time. While at any given time you're limited to a thousand posts, nothing stops you from growing your own database over the course of months. It's worth noting the following script might take about an hour to complete. If you don't feel like waiting, feel free to proceed and use the downloadable SQLite file. Also, if you run it now you are not likely to get the exact same output as when it was first run to create the output shown in this chapter.

Let's look at our data retrieval function, as shown in the following listing.

> **Listing 8.2   Reddit data retrieval and storage in SQLite**

Specific fields of the topic are appended to the list. We only use the title and text throughout the exercise but the topic ID would be useful for building your own (bigger) database of topics.

From subreddits, get hottest 1,000 (in our case) topics.

```
def prawGetData(limit,subredditName):
    topics = r.get_subreddit(subredditName).get_hot(limit=limit)
    commentInsert = []
    topicInsert = []
    topicNBR = 1
    for topic in topics:
        if (float(topicNBR)/limit)*100 in xrange(1,100):
            print '********** TOPIC:' + str(topic.id)
+ ' ********COMPLETE: ' + str((float(topicNBR)/limit)*100)
+ ' % ****'
        topicNBR += 1
        try:
            topicInsert.append((topic.title,topic.selftext,topic.id,
            subredditName))
        except:
            pass
        try:
            for comment in topic.comments:
                commentInsert.append((comment.body,comment.id,
topic.title,topic.selftext,topic.id,subredditName))
        except:
            pass
    print '*****************************'
    print 'INSERTING DATA INTO SQLITE'
    c.executemany('INSERT INTO topics VALUES (?,?,?,?)', topicInsert)
    print 'INSERTED TOPICS'
    c.executemany('INSERT INTO comments VALUES (?,?,?,?,?,?)', commentInsert)
    print 'INSERTED COMMENTS'
    conn.commit()

for subject in subreddits:
    prawGetData(limit=limit,subredditName=subject)
```

This part is an informative print and not necessary for code to work. It only informs you about the download progress.

Append comments to a list. These are not used in the exercise but now you have them for experimentation.

Insert all topics into SQLite database.

Insert all comments into SQLite database.

Commit changes (data insertions) to database. Without the commit, no data will be inserted.

The function is executed for all subreddits we specified earlier.

The `prawGetData()` function retrieves the "hottest" topics in its subreddit, appends this to an array, and then gets all its related comments. This goes on until a thousand topics are reached or no more topics exist to fetch and everything is stored in the SQLite database. The print statements are there to inform you on its progress toward gathering a thousand topics. All that's left for us to do is execute the function for each subreddit.

If you'd like this analysis to incorporate more than two subreddits, this is a matter of adding an extra category to the subreddits array.

With the data collected, we're ready to move on to data preparation.

8.3.4 Step 3: Data preparation

As always, data preparation is the most crucial step to get correct results. For text mining this is even truer since we don't even start off with structured data.

The upcoming code is available online as IPython file "Chapter 8 data preparation and analysis." Let's start by importing the required libraries and preparing the SQLite database, as shown in the following listing.

Listing 8.3   Text mining, libraries, corpora dependencies, and SQLite database connection

```
import sqlite3
import nltk                                    Import all
import matplotlib.pyplot as plt                required
from collections import OrderedDict            libraries
import random

nltk.download('punkt')            Download corpora
nltk.download('stopwords')        we make use of

conn = sqlite3.connect('reddit.db')      Make a connection to SQLite database
c = conn.cursor()                        that contains our Reddit data
```

In case you haven't already downloaded the full NLTK corpus, we'll now download the part of it we'll use. Don't worry if you already downloaded it, the script will detect if your corpora is up to date.

Our data is still stored in the Reddit SQLite file so let's create a connection to it.

Even before exploring our data we know of at least two things we have to do to clean the data: stop word filtering and lowercasing.

A general word filter function will help us filter out the unclean parts. Let's create one in the following listing.

Listing 8.4   Word filtering and lowercasing functions

```
def wordFilter(excluded,wordrow):
    filtered = [word for word in wordrow if word not in excluded]
    return filtered
stopwords = nltk.corpus.stopwords.words('english')
def lowerCaseArray(wordrow):
    lowercased = [word.lower() for word in wordrow]
    return lowercased
```

wordFilter() function will remove a term from an array of terms

Stop word variable contains English stop words per default present in NLTK

lowerCaseArray() function transforms any term to its lowercased version

The English stop words will be the first to leave our data. The following code will provide us these stop words:

```
stopwords = nltk.corpus.stopwords.words('english')
print stopwords
```

Figure 8.14 shows the list of English stop words in NLTK.

```
stopwords = nltk.corpus.stopwords.words('english')
print stopwords
```
```
[u'i', u'me', u'my', u'myself', u'we', u'our', u'ours', u'ourselves', u'you',
u'your', u'yours', u'yourself', u'yourselves', u'he', u'him', u'his', u'himsel
f', u'she', u'her', u'hers', u'herself', u'it', u'its', u'itself', u'they', u'th
em', u'their', u'theirs', u'themselves', u'what', u'which', u'who', u'whom', u't
his', u'that', u'these', u'those', u'am', u'is', u'are', u'was', u'were', u'be',
u'been', u'being', u'have', u'has', u'had', u'having', u'do', u'does', u'did',
u'doing', u'a', u'an', u'the', u'and', u'but', u'if', u'or', u'because', u'as',
u'until', u'while', u'of', u'at', u'by', u'for', u'with', u'about', u'against',
u'between', u'into', u'through', u'during', u'before', u'after', u'above', u'bel
ow', u'to', u'from', u'up', u'down', u'in', u'out', u'on', u'off', u'over', u'un
der', u'again', u'further', u'then', u'once', u'here', u'there', u'when', u'wher
e', u'why', u'how', u'all', u'any', u'both', u'each', u'few', u'more', u'most',
u'other', u'some', u'such', u'no', u'nor', u'not', u'only', u'own', u'same', u's
o', u'than', u'too', u'very', u's', u't', u'can', u'will', u'just', u'don', u'sh
ould', u'now']
```

Figure 8.14   English stop words list in NLTK

With all the necessary components in place, let's have a look at our first data processing function in the following listing.

Listing 8.5 First data preparation function and execution

We'll use data['all_words']
for data exploration.

Create pointer
to AWLite data.

row[0] is
title, row[1]
is topic text;
we turn them
into a single
text blob.

Fetch data
row by row.

```
def data_processing(sql):
    c.execute(sql)
    data = {'wordMatrix':[],'all_words':[]}
    row = c.fetchone()
    while row is not None:
        wordrow = nltk.tokenize.word_tokenize(row[0]+" "+row[1])
        wordrow_lowercased = lowerCaseArray(wordrow)
        wordrow_nostopwords = wordFilter(stopwords,wordrow_lowercased)
        data['all_words'].extend(wordrow_nostopwords)
        data['wordMatrix'].append(wordrow_nostopwords)
        row = c.fetchone()
    return data

subreddits = ['datascience','gameofthrones']
data = {}
for subject in subreddits:
    data[subject] = data_processing(sql='''SELECT
     topicTitle,topicText,topicCategory FROM topics
WHERE topicCategory = '''+"'"+subject+"'")
```

Get new document
from SQLite database.

Our subreddits as
defined earlier.

Call data processing
function for every
subreddit.

data['wordMatrix'] is a matrix
comprised of word vectors;
1 vector per document.

Our data_processing() function takes in a SQL statement and returns the document-term matrix. It does this by looping through the data one entry (Reddit topic) at a time and combines the topic title and topic body text into a single word vector with the use of word tokenization. A *tokenizer* is a text handling script that cuts the text into pieces. You have many different ways to tokenize a text: you can divide it into sentences or words, you can split by space and punctuations, or you can take other characters into account, and so on. Here we opted for the standard NLTK word tokenizer. This word tokenizer is simple; all it does is split the text into terms if there's a space between the words. We then lowercase the vector and filter out the stop words. Note how the order is important here; a stop word in the beginning of a sentence wouldn't be filtered if we first filter the stop words before lowercasing. For instance in "I like Game of Thrones," the "I" would not be lowercased and thus would not be filtered out. We then create a word matrix (term-document matrix) and a list containing all the words. Notice how we extend the list without filtering for doubles; this way we can create a histogram on word occurrences during data exploration. Let's execute the function for our two topic categories.

Figure 8.15 shows the first word vector of the "datascience" category.

```
print data['datascience']['wordMatrix'][0]
```

```
print data['datascience']['wordMatrix'][0]
```

```
[u'data', u'science', u'freelancing', u"'m", u'currently', u'master
s', u'program', u'studying', u'business', u'analytics', u"'m", u'try
ing', u'get', u'data', u'freelancing', u'.', u"'m", u'still', u'lear
ning', u'skill', u'set', u'typically', u'see', u'right', u"'m", u'fa
irly', u'proficient', u'sql', u'know', u'bit', u'r.', u'freelancer
s', u'find', u'jobs', u'?']
```

Figure 8.15 The first word vector of the "datascience" category after first data processing attempt

This sure looks polluted: punctuations are kept as separate terms and several words haven't even been split. Further data exploration should clarify a few things for us.

8.3.5 *Step 4: Data exploration*

We now have all our terms separated, but the sheer size of the data hinders us from getting a good grip on whether it's clean enough for actual use. By looking at a single vector, we already spot a few problems though: several words haven't been split correctly and the vector contains many single-character terms. Single character terms might be good topic differentiators in certain cases. For example, an economic text will contain more \$, £, and € signs than a medical text. But in most cases these one-character terms are useless. First, let's have a look at the frequency distribution of our terms.

```
wordfreqs_cat1 = nltk.FreqDist(data['datascience']['all_words'])
plt.hist(wordfreqs_cat1.values(), bins = range(10))
plt.show()
wordfreqs_cat2 = nltk.FreqDist(data['gameofthrones']['all_words'])
plt.hist(wordfreqs_cat2.values(), bins = range(20))
plt.show()
```

By drawing a histogram of the frequency distribution (figure 8.16) we quickly notice that the bulk of our terms only occur in a single document.

Single-occurrence terms such as these are called *hapaxes*, and model-wise they're useless because a single occurrence of a feature is never enough to build a reliable model. This is good news for us; cutting these hapaxes out will significantly shrink our data without harming our eventual model. Let's look at a few of these single-occurrence terms.

```
print wordfreqs_cat1.hapaxes()
print wordfreqs_cat2.hapaxes()
```

Terms we see in figure 8.17 make sense, and if we had more data they'd likely occur more often.

```
print wordfreqs_cat1.hapaxes()
print wordfreqs_cat2.hapaxes()
```

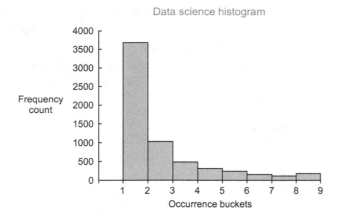

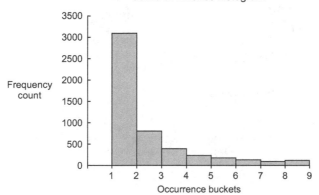

Figure 8.16 This histogram of term frequencies shows both the "data science" and "game of thrones" term matrices have more than 3,000 terms that occur once.

Least frequent terms within data science posts

```
print wordfreqs_cat1.hapaxes()
```

[u'post-grad', u'marching', u'cytoscape', u'wizardry', u"'pure", u'i
mmature', u'socrata', u'filenotfoundexception', u'side-by-side', u'b
ringing', u'non-experienced', u'zestimate', u'formatting*', u'sustai

Least frequent terms within Game of Thrones posts

```
print wordfreqs_cat2.hapaxes()
```

[u'hordes', u'woods', u'comically', u'pack', u'seventy-seven', u"'co
ntext", u'shaving', u'kennels', u'differently', u'screaming', u'her-
', u'complainers', u'sailed', u'contributed', u'payoff', u'hallucina

Figure 8.17 "Data science" and "game of thrones" single occurrence terms (hapaxes)

Many of these terms are incorrect spellings of otherwise useful ones, such as: Jaimie is Jaime (Lannister), Milisandre would be Melisandre, and so on. A decent *Game of Thrones*-specific thesaurus could help us find and replace these misspellings with a fuzzy search algorithm. This proves data cleaning in text mining can go on indefinitely if you so desire; keeping effort and payoff in balance is crucial here.

Let's now have a look at the most frequent words.

```
print wordfreqs_cat1.most_common(20)
print wordfreqs_cat2.most_common(20)
```

Figure 8.18 shows the output of asking for the top 20 most common words for each category.

Most frequent words within data science posts

```
print wordfreqs_cat1.most_common(20)
```

```
[(u'.', 2833), (u',', 2831), (u'data', 1882), (u'?', 1190), (u'scien
ce', 887), (u')', 812), (u'(', 739), (u"'m", 566), (u':', 548), (u'w
ould', 427), (u"'s", 323), (u'like', 321), (u"n't", 288), (u'get', 2
52), (u'know', 225), (u"'ve", 213), (u'scientist', 211), (u'!', 20
9), (u'work', 204), (u'job', 199)]
```

Most frequent words within Game of Thrones posts

```
print wordfreqs_cat2.most_common(20)
```

```
[(u'.', 2909), (u',', 2478), (u'[', 1422), (u']', 1420), (u'?', 113
9), (u"'s", 886), (u"n't", 494), (u')', 452), (u'(', 426), (u's5', 3
99), (u':', 380), (u'spoilers', 332), (u'show', 325), (u'would', 31
1), (u"''", 305), (u'``', 276), (u'think', 248), (u'season', 244),
(u'like', 243), (u'one', 238)]
```

Figure 8.18   Top 20 most frequent words for the "data science" and "game of thrones" posts

Now this looks encouraging: several common words do seem specific to their topics. Words such as "data," "science," and "season" are likely to become good differentiators. Another important thing to notice is the abundance of the single character terms such as "." and ","; we'll get rid of these.

With this extra knowledge, let's revise our data preparation script.

8.3.6   *Step 3 revisited: Data preparation adapted*

This short data exploration has already drawn our attention to a few obvious tweaks we can make to improve our text. Another important one is stemming the terms.

The following listing shows a simple stemming algorithm called "snowball stemming." These snowball stemmers can be language-specific, so we'll use the English one; however, it does support many languages.

Listing 8.6 The Reddit data processing revised after data exploration

```
stemmer = nltk.SnowballStemmer("english")                    ◄──── Initializes stemmer
def wordStemmer(wordrow):                                           from NLTK library.
    stemmed = [stemmer.stem(word) for word in wordrow]
    return stemmed

manual_stopwords = [',','.',')',',',',','(','m',"'m","n't",'e.g',"'ve",'s','#','/
    ',''`'',"'s","'''",'!','r',']',',','=','[','s','&','%','*','...','1','2','3','
    4','5','6','7','8','9','10','--',"'",';','-',':']                ◄──── Stop words array
                                                                          defines terms to
                                                                          remove/ignore.
def data_processing(sql,manual_stopwords):                   ◄──┐
    #create pointer to the sqlite data                          │  Now we define
    c.execute(sql)                                              │  our revised data
    data = {'wordMatrix':[],'all_words':[]}                     │  preparation.
    interWordMatrix = []
    interWordList = []


    row = c.fetchone()                                       ◄──┤ Fetch data (reddit posts) one
    while row is not None:                                        │ by one from SQLite database.
        tokenizer = nltk.tokenize.RegexpTokenizer(r'\w+|[^\w\s]+')

        wordrow = tokenizer.tokenize(row[0]+" "+row[1])
        wordrow_lowercased = lowerCaseArray(wordrow)
        wordrow_nostopwords = wordFilter(stopwords,wordrow_lowercased)

        wordrow_nostopwords =                                     ◄──┐ Temporary word
            wordFilter(manual_stopwords,wordrow_nostopwords)          │ list used to
        wordrow_stemmed = wordStemmer(wordrow_nostopwords)            │ remove hapaxes
                                                                      │ later on.
        interWordList.extend(wordrow_stemmed)                 ◄──┘
        interWordMatrix.append(wordrow_stemmed)

        row = c.fetchone()                                    ◄──── Get new
                                                                    topic.
    wordfreqs = nltk.FreqDist(interWordList) s                ◄──── Make frequency
    hapaxes = wordfreqs.hapaxes()                             ◄──── distribution of
    for wordvector in interWordMatrix:                              all terms.
        wordvector_nohapexes = wordFilter(hapaxes,wordvector) ◄──── Get list of hapaxes.
        data['wordMatrix'].append(wordvector_nohapexes)
        data['all_words'].extend(wordvector_nohapexes)        ◄──── Remove hapaxes in
                                                                    each word vector.
    return data

for subject in subreddits:                                    ◄──── Extend list of all
    data[subject] = data_processing(sql='''SELECT                   terms with corrected
     topicTitle,topicText,topicCategory FROM topics                 word vector.
WHERE topicCategory = '''+"'"+subject+"'",
manual_stopwords=manual_stopwords)                             ◄──── Run new data
                                                                    processing function
                                                                    for both subreddits.
```

row[0] and row[1] contain the title and text of the post, respectively. We combine them into a single text blob.

Remove manually added stop words from text blob.

Temporary word matrix; will become final word matrix after hapaxes removal.

Loop through temporary word matrix.

Append correct word vector to final word matrix.

Notice the changes since the last `data_processing()` function. Our tokenizer is now a regular expression tokenizer. Regular expressions are not part of this book and are often considered challenging to master, but all this simple one does is cut the text into words. For words, any alphanumeric combination is allowed (\w), so there are no more special characters or punctuations. We also applied the word stemmer and removed a list of extra stop words. And, all the hapaxes are removed at the end because everything needs to be stemmed first. Let's run our data preparation again.

If we did the same exploratory analysis as before, we'd see it makes more sense, and we have no more hapaxes.

```
print wordfreqs_cat1.hapaxes()
print wordfreqs_cat2.hapaxes()
```

Let's take the top 20 words of each category again (see figure 8.19).

Top 20 most common "Data Science" terms after more intense data cleansing

```
wordfreqs_cat1 = nltk.FreqDist(data['datascience']['all_words'])
print wordfreqs_cat1.most_common(20)
```

```
[(u'data', 1971), (u'scienc', 955), (u'would', 418), (u'work', 368), (u'use', 34
7), (u'program', 343), (u'learn', 342), (u'like', 341), (u'get', 325), (u'scient
ist', 310), (u'job', 268), (u'cours', 265), (u'look', 257), (u'know', 239), (u's
tatist', 228), (u'want', 225), (u've', 223), (u'python', 205), (u'year', 204),
(u'time', 196)]
```

Top 20 most common "Game of Thrones" terms after more intense data cleansing

```
wordfreqs_cat2 = nltk.FreqDist(data['gameofthrones']['all_words'])
print wordfreqs_cat2.most_common(20)
```

```
[(u's5', 426), (u'spoiler', 374), (u'show', 362), (u'episod', 300), (u'think', 2
89), (u'would', 287), (u'season', 286), (u'like', 282), (u'book', 271), (u'one',
249), (u'get', 236), (u'sansa', 232), (u'scene', 216), (u'cersei', 213), (u'kno
w', 192), (u'go', 188), (u'king', 183), (u'throne', 181), (u'see', 177), (u'char
act', 177)]
```

Figure 8.19   Top 20 most frequent words in "data science" and "game of thrones" Reddit posts after data preparation

We can see in figure 8.19 how the data quality has improved remarkably. Also, notice how certain words are shortened because of the stemming we applied. For instance, "science" and "sciences" have become "scienc;" "courses" and "course" have become "cours," and so on. The resulting terms are not actual words but still interpretable. If you insist on your terms remaining actual words, lemmatization would be the way to go.

With the data cleaning process "completed" (remark: a text mining cleansing exercise can almost never be fully completed), all that remains is a few data transformations to get the data in the bag of words format.

First, let's label all our data and also create a holdout sample of 100 observations per category, as shown in the following listing.

Listing 8.7 Final data transformation and data splitting before modeling

Holdout sample is comprised of unlabeled data from the two subreddits: 100 observations from each data set. The labels are kept in a separate data set.

Holdout sample will be used to determine the model's flaws by constructing a confusion matrix.

```
holdoutLength  = 100
```

```
labeled_data1 = [(word,'datascience') for word in
     data['datascience']['wordMatrix'][holdoutLength:]]
labeled_data2 = [(word,'gameofthrones') for word in
     data['gameofthrones']['wordMatrix'][holdoutLength:]]
labeled_data = []
labeled_data.extend(labeled_data1)
labeled_data.extend(labeled_data2)
```

We create a single data set with every word vector tagged as being either 'datascience' or 'gameofthrones.' We keep part of the data aside for holdout sample.

```
holdout_data = data['datascience']['wordMatrix'][:holdoutLength]
holdout_data.extend(data['gameofthrones']['wordMatrix'][:holdoutLength])
holdout_data_labels = ([('datascience')
for _ in xrange(holdoutLength)] + [('gameofthrones') for _ in
     xrange(holdoutLength)])
```

```
data['datascience']['all_words_dedup'] =
list(OrderedDict.fromkeys(
data['datascience']['all_words']))
data['gameofthrones']['all_words_dedup'] =
list(OrderedDict.fromkeys(
data['gameofthrones']['all_words']))
all_words = []
all_words.extend(data['datascience']['all_words_dedup'])
all_words.extend(data['gameofthrones']['all_words_dedup'])
all_words_dedup = list(OrderedDict.fromkeys(all_words))
```

A list of all unique terms is created to build the bag of words data we need for training or scoring a model.

Data for model training and testing is first shuffled.

```
prepared_data = [({word: (word in x[0]) for word
in all_words_dedup}, x[1]) for x in labeled_data]
prepared_holdout_data = [({word: (word in x[0])
for word in all_words_dedup})
for x in holdout_data]
```

Data is turned into a binary bag of words format.

```
random.shuffle(prepared_data)
train_size = int(len(prepared_data) * 0.75)
train = prepared_data[:train_size]
test = prepared_data[train_size:]
```

Size of training data will be 75% of total and remaining 25% will be used for testing model performance.

The holdout sample will be used for our final test of the model and the creation of a confusion matrix. A *confusion matrix* is a way of checking how well a model did on previously unseen data. The matrix shows how many observations were correctly and incorrectly classified.

Before creating or training and testing data we need to take one last step: pouring the data into a bag of words format where every term is given either a "True" or "False" label depending on its presence in that particular post. We also need to do this for the unlabeled holdout sample.

Our prepared data now contains every term for each vector, as shown in figure 8.20.

```
print prepared_data[0]
```

```
print prepared_data[0]

({u'sunspear': False, u'profici': False, u'pardon': False, u'selye
s': False, u'four': False, u'davo': False, u'sleev': False, u'slee
                            ⋮
u'daeron': False, u'portion': False, u'emerg': False, u'fifti': Fals
e, u'decemb': False, u'defend': False, u'sincer': False}, 'datascien
ce')
```

Figure 8.20   A binary bag of words ready for modeling is very sparse data.

We created a big but sparse matrix, allowing us to apply techniques from chapter 5 if it was too big to handle on our machine. With such a small table, however, there's no need for that now and we can proceed to shuffle and split the data into a training and test set.

While the biggest part of your data should always go to the model training, an optimal split ratio exists. Here we opted for a 3-1 split, but feel free to play with this. The more observations you have, the more freedom you have here. If you have few observations you'll need to allocate relatively more to training the model. We're now ready to move on to the most rewarding part: data analysis.

8.3.7   *Step 5: Data analysis*

For our analysis we'll fit two classification algorithms to our data: Naïve Bayes and decision trees. Naïve Bayes was explained in chapter 3 and decision tree earlier in this chapter.

Let's first test the performance of our Naïve Bayes classifier. NLTK comes with a classifier, but feel free to use algorithms from other packages such as SciPy.

```
classifier  = nltk.NaiveBayesClassifier.train(train)
```

With the classifier trained we can use the test data to get a measure on overall accuracy.

```
nltk.classify.accuracy(classifier, test)
```

```
nltk.classify.accuracy(classifier, test)

0.9681528662420382
```

Figure 8.21 **Classification accuracy is a measure representing what percentage of observations was correctly classified on the test data.**

The accuracy on the test data is estimated to be greater than 90%, as seen in figure 8.21. *Classification accuracy* is the number of correctly classified observations as a percentage of the total number of observations. Be advised, though, that this can be different in your case if you used different data.

```
nltk.classify.accuracy(classifier, test)
```

That's a good number. We can now lean back and relax, right? No, not really. Let's test it again on the 200 observations holdout sample and this time create a confusion matrix.

```
classified_data = classifier.classify_many(prepared_holdout_data)
cm = nltk.ConfusionMatrix(holdout_data_labels, classified_data)
print cm
```

The confusion matrix in figure 8.22 shows us the 97% is probably over the top because we have 28 (23 + 5) misclassified cases. Again, this can be different with your data if you filled the SQLite file yourself.

```
              |       g |
              |       a |
              |   d   m |
              |   a   e |
              |   t   o |
              |   a   f |
              |   s   t |
              |   c   h |
              |   i   r |
              |   e   o |
              |   n   n |
              |   c   e |
              |   e   s |
--------------+-------+
   datascience |<77>23 |
 gameofthrones |  5<95>|
--------------+-------+
(row = reference; col = test)
```

Figure 8.22 **Naïve Bayes model confusion matrix shows 28 (23 + 5) observations out of 200 were misclassified**

Twenty-eight misclassifications means we have an 86% accuracy on the holdout sample. This needs to be compared to randomly assigning a new post to either the "datascience" or "gameofthrones" group. If we'd randomly assigned them, we could expect an accuracy of 50%, and our model seems to perform better than that. Let's look at what it uses to determine the categories by digging into the most informative model features.

```
print(classifier.show_most_informative_features(20))
```

Figure 8.23 shows the top 20 terms capable of distinguishing between the two categories.

```
Most Informative Features
                 data = True          datasc : gameof =    365.1 : 1.0
                scene = True          gameof : datasc =     63.8 : 1.0
               season = True          gameof : datasc =     62.4 : 1.0
                 king = True          gameof : datasc =     47.6 : 1.0
                   tv = True          gameof : datasc =     45.1 : 1.0
                 kill = True          gameof : datasc =     31.5 : 1.0
              compani = True          datasc : gameof =     28.5 : 1.0
               analysi = True         datasc : gameof =     27.1 : 1.0
              process = True          datasc : gameof =     25.5 : 1.0
                appli = True          datasc : gameof =     25.5 : 1.0
             research = True          datasc : gameof =     23.2 : 1.0
               episod = True          gameof : datasc =     22.2 : 1.0
               market = True          datasc : gameof =     21.7 : 1.0
                watch = True          gameof : datasc =     21.6 : 1.0
                  man = True          gameof : datasc =     21.0 : 1.0
                north = True          gameof : datasc =     20.8 : 1.0
                   hi = True          datasc : gameof =     20.4 : 1.0
                level = True          datasc : gameof =     19.1 : 1.0
                learn = True          datasc : gameof =     16.9 : 1.0
                  job = True          datasc : gameof =     16.6 : 1.0
```

Figure 8.23   The most important terms in the Naïve Bayes classification model

The term "data" is given heavy weight and seems to be the most important indicator of whether a topic belongs in the data science category. Terms such as "scene," "season," "king," "tv," and "kill" are good indications the topic is *Game of Thrones* rather than data science. All these things make perfect sense, so the model passed both the accuracy and the sanity check.

The Naïve Bayes does well, so let's have a look at the decision tree in the following listing.

Listing 8.8   Decision tree model training and evaluation

Create confusion matrix based on classification results and actual labels ⟶ Train decision tree classifier ⟶ Test classifier accuracy

```
classifier2 = nltk.DecisionTreeClassifier.train(train)
nltk.classify.accuracy(classifier2, test)
classified_data2 = classifier2.classify_many(prepared_holdout_data)
cm = nltk.ConfusionMatrix(holdout_data_labels, classified_data2)
print cm
```

Show confusion matrix

Attempt to classify holdout data (scoring)

```
nltk.classify.accuracy(classifier2, test)
0.9333333333333333
```

Figure 8.24 Decision tree model accuracy

As shown in figure 8.24, the promised accuracy is 93%.

We now know better than to rely solely on this single test, so once again we turn to a confusion matrix on a second set of data, as shown in figure 8.25.

Figure 8.25 shows a different story. On these 200 observations of the holdout sample the decision tree model tends to classify well when the post is about *Game of Thrones* but fails miserably when confronted with the data science posts. It seems the model has a preference for *Game of Thrones*, and can you blame it? Let's have a look at the actual model, even though in this case we'll use the Naïve Bayes as our final model.

```
print(classifier2.pseudocode(depth=4))
```

```
              | g |
              | a |
              | m |
          | d | e |
          | a | o |
          | t | f |
          | a | t |
          | s | h |
          | c | r |
          | i | o |
          | e | n |
          | n | e |
          | c | s |
          | e |   |
-------------+------+
datascience |<26>74 |
gameofthrones |  2<98>|
-------------+------+
(row = reference; col = test)
```

Figure 8.25 Confusion matrix on decision tree model

The decision tree has, as the name suggests, a tree-like model, as shown in figure 8.26.

The Naïve Bayes considers all the terms and has weights attributed, but the decision tree model goes through them sequentially, following the path from the root to the outer branches and leaves. Figure 8.26 only shows the top four layers, starting with the term "data." If "data" is present in the post, it's always data science. If "data" can't be found, it checks for the term "learn," and so it continues. A possible reason why this decision tree isn't performing well is the lack of pruning. When a decision tree is built it has many leaves, often too many. A tree is then pruned to a certain level to minimize overfitting. A big advantage of decision trees is the implicit interaction effects between words it

```
if data == False:
  if learn == False:
    if python == False:
      if tool == False: return 'gameofthrones'
      if tool == True: return 'datascience'
    if python == True: return 'datascience'
  if learn == True:
    if go == False:
      if wrong == False: return 'datascience'
      if wrong == True: return 'gameofthrones'
    if go == True:
      if upload == False: return 'gameofthrones'
      if upload == True: return 'datascience'
if data == True: return 'datascience'
```

Figure 8.26 Decision tree model tree structure representation

takes into account when constructing the branches. When multiple terms together create a stronger classification than single terms, the decision tree will actually outperform the Naïve Bayes. We won't go into the details of that here, but consider this one of the next steps you could take to improve the model.

We now have two classification models that give us insight into how the two contents of the subreddits differ. The last step would be to share this newfound information with other people.

8.3.8   *Step 6: Presentation and automation*

As a last step we need to use what we learned and either turn it into a useful application or present our results to others. The last chapter of this book discusses building an interactive application, as this is a project in itself. For now we'll content ourselves with a nice way to convey our findings. A nice graph or, better yet, an interactive graph, can catch the eye; it's the icing on the presentation cake. While it's easy and tempting to represent the numbers as such or a bar chart at most, it could be nice to go one step further.

For instance, to represent the Naïve Bayes model, we could use a force graph (figure 8.27), where the bubble and link size represent how strongly related a word is to the "game of thrones" or "data science" subreddits. Notice how the words on the bubbles are often cut off; remember this is because of the stemming we applied.

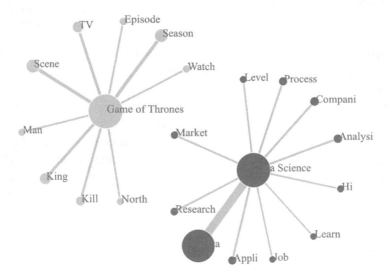

Figure 8.27   **Interactive force graph with the top 20 Naïve Bayes significant terms and their weights**

While figure 8.27 in itself is static, you can open the HTML file "forceGraph.html" to enjoy the d3.js force graph effect as explained earlier in this chapter. d3.js is outside of this book's scope but you don't need an elaborate knowledge of d3.js to use it. An extensive set of examples can be used with minimal adjustments to the code provided at https://github.com/mbostock/d3/wiki/Gallery. All you need is common sense and a minor knowledge of JavaScript. The code for the force graph example can found at http://bl.ocks.org/mbostock/4062045.

We can also represent our decision tree in a rather original way. We could go for a fancy version of an actual tree diagram, but the following sunburst diagram is more original and equally fun to use.

Figure 8.28 shows the top layer of the sunburst diagram. It's possible to zoom in by clicking a circle segment. You can zoom back out by clicking the center circle. The code for this example can be found at http://bl.ocks.org/metmajer/5480307.

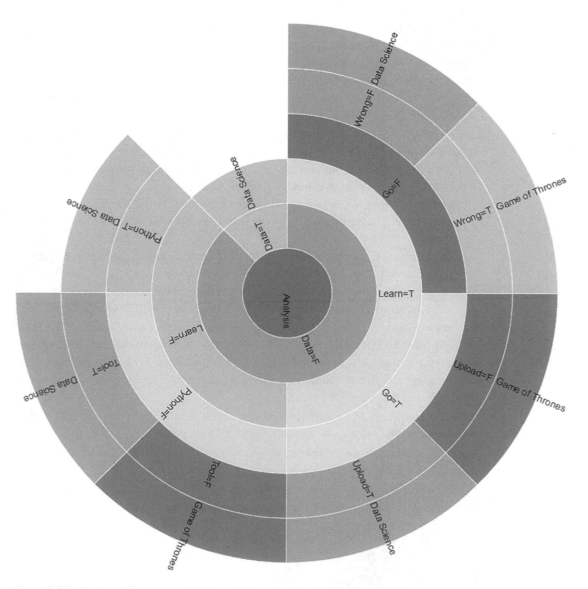

Figure 8.28   Sunburst diagram created from the top four branches of the decision tree model

Showing your results in an original way can be key to a successful project. People never appreciate the effort you've put into achieving your results if you can't communicate them and they're meaningful to them. An original data visualization here and there certainly helps with this.

8.4   *Summary*

- Text mining is widely used for things such as entity identification, plagiarism detection, topic identification, translation, fraud detection, spam filtering, and more.
- Python has a mature toolkit for text mining called NLTK, or the natural language toolkit. NLTK is good for playing around and learning the ropes; for real-life applications, however, Scikit-learn is usually considered more "production-ready." Scikit-learn is extensively used in previous chapters.
- The data preparation of textual data is more intensive than numerical data preparation and involves extra techniques, such as
  - *Stemming*—Cutting the end of a word in a smart way so it can be matched with some conjugated or plural versions of this word.
  - *Lemmatization*—Like stemming, it's meant to remove doubles, but unlike stemming, it looks at the meaning of the word.
  - *Stop word filtering*—Certain words occur too often to be useful and filtering them out can significantly improve models. Stop words are often corpus-specific.
  - *Tokenization*—Cutting text into pieces. Tokens can be single words, combinations of words (n-grams), or even whole sentences.
  - *POS Tagging*—Part-of-speech tagging. Sometimes it can be useful to know what the function of a certain word within a sentence is to understand it better.
- In our case study we attempted to distinguish Reddit posts on "Game of Thrones" versus posts on "data science." In this endeavor we tried both the Naïve Bayes and decision tree classifiers. Naïve Bayes assumes all features to be independent of one another; the decision tree classifier assumes dependency, allowing for different models.
- In our example, Naïve Bayes yielded the better model, but very often the decision tree classifier does a better job, usually when more data is available.
- We determined the performance difference using a confusion matrix we calculated after applying both models on new (but labeled) data.
- When presenting findings to other people, it can help to include an interesting data visualization capable of conveying your results in a memorable way.

Data visualization
to the end user

This chapter covers

- Considering options for data visualization for your end users
- Setting up a basic Crossfilter MapReduce application
- Creating a dashboard with dc.js
- Working with dashboard development tools

APPLICATION FOCUSED CHAPTER You'll notice quickly this chapter is certainly different from chapters 3 to 8 in that the focus here lies on step 6 of the data science process. More specifically, what we want to do here is create a small data science application. Therefore, we won't follow the data science process steps here. The data used in the case study is only partly real but functions as data flowing from either the data preparation or data modeling stage. Enjoy the ride.

Often, data scientists must deliver their new insights to the end user. The results can be communicated in several ways:

- *A one-time presentation*—Research questions are one-shot deals because the business decision derived from them will bind the organization to a certain course for many years to come. Take, for example, company investment decisions: *Do we distribute our goods from two distribution centers or only one? Where do they need to be located for optimal efficiency?* When the decision is made, the exercise may not be repeated until you've retired. In this case, the results are delivered as a report with a presentation as the icing on the cake.

■ *A new viewport on your data*—The most obvious example here is customer segmentation. Sure, the segments themselves will be communicated via reports and presentations, but in essence they form tools, not the end result itself. When a clear and relevant customer segmentation is discovered, it can be fed back to the database as a new dimension on the data from which it was derived. From then on, people can make their own reports, such as how many products were sold to each segment of customers.

■ *A real-time dashboard*—Sometimes your task as a data scientist doesn't end when you've discovered the new information you were looking for. You can send your information back to the database and be done with it. But when other people start making reports on this newly discovered gold nugget, they might interpret it incorrectly and make reports that don't make sense. As the data scientist who discovered this new information, you must set the example: make the first refreshable report so others, mainly reporters and IT, can understand it and follow in your footsteps. Making the first dashboard is also a way to shorten the delivery time of your insights to the end user who wants to use it on an everyday basis. This way, at least they already have something to work with until the reporting department finds the time to create a permanent report on the company's reporting software.

You might have noticed that a few important factors are at play:

■ What *kind of decision* are you supporting? Is it a strategic or an operational one? Strategic decisions often only require you to analyze and report once, whereas operational decisions require the report to be refreshed regularly.

■ *How big is your organization?* In smaller ones you'll be in charge of the entire cycle: from data gathering to reporting. In bigger ones a team of reporters might be available to make the dashboards for you. But even in this last situation, delivering a prototype dashboard can be beneficial because it presents an example and often shortens delivery time.

Although the entire book is dedicated to generating insights, in this last chapter we'll focus on delivering an operational dashboard. Creating a presentation to promote your findings or presenting strategic insights is out of the scope of this book.

9.1   *Data visualization options*

You have several options for delivering a dashboard to your end users. Here we'll focus on a single option, and by the end of this chapter you'll be able to create a dashboard yourself.

This chapter's case is that of a hospital pharmacy with a stock of a few thousand medicines. The government came out with a new norm to all pharmacies: all medicines should be checked for their sensitivity to light and be stored in new, special containers. One thing the government didn't supply to the pharmacies was an actual list of light-sensitive medicines. This is no problem for you as a data scientist because every medicine has a patient information leaflet that contains this information. You distill the information with the clever use of text mining and assign a "light sensitive" or "not light sensitive" tag to each medicine. This information is then uploaded to the central database. In addition, the pharmacy needs to know how many containers would be necessary. For this they give you access to the pharmacy stock data. When you draw a sample with only the variables you require, the data set looks like figure 9.1 when opened in Excel.

	A	B	C	D	E	F
1	MedName	LightSen	Date	StockOut	StockIn	Stock
2	Acupan 30 mg	No	1/01/2015	-8	150	142
3	Acupan 30 mg	No	2/01/2015	-6	5	141
4	Acupan 30 mg	No	3/01/2015	-2	0	139
5	Acupan 30 mg	No	4/01/2015	0	5	144
6	Acupan 30 mg	No	5/01/2015	-8	0	136
7	Acupan 30 mg	No	6/01/2015	-1	0	135
8	Acupan 30 mg	No	7/01/2015	-1	15	149
9	Acupan 30 mg	No	8/01/2015	-10	10	149
10	Acupan 30 mg	No	9/01/2015	-8	15	156

Figure 9.1 Pharmacy medicines data set opened in Excel: the first 10 lines of stock data are enhanced with a light-sensitivity variable

As you can see, the information is time-series data for an entire year of stock movement, so every medicine thus has 365 entries in the data set. Although the case study is an existing one and the medicines in the data set are real, the values of the other variables presented here were randomly generated, as the original data is classified. Also, the data set is limited to 29 medicines, a little more than 10,000 lines of data. Even though people do create reports using crossfilter.js (a Javascript MapReduce library) and dc.js (a Javascript dashboarding library) with more than a million lines of data, for the example's sake you'll use a fraction of this amount. Also, it's not recommended to load your entire database into the user's browser; the browser will freeze while loading, and if it's too much data, the browser will even crash. Normally data is precalculated on the server and parts of it are requested using, for example, a REST service.

To turn this data into an actual dashboard you have many options and you can find a short overview of the tools later in this chapter.

Among all the options, for this book we decided to go with dc.js, which is a crossbreed between the JavaScript MapReduce library Crossfilter and the data visualization library d3.js. Crossfilter was developed by Square Register, a company that handles payment transactions; it's comparable to PayPal but its focus is on mobile. Square developed Crossfilter to allow their customers extremely speedy slice and dice on their payment history. Crossfilter is not the only JavaScript library capable of Map-Reduce processing, but it most certainly does the job, is open source, is free to use, and is maintained by an established company (Square). Example alternatives to Crossfilter are Map.js, Meguro, and Underscore.js. JavaScript might not be known as a data crunching language, but these libraries do give web browsers that extra bit of punch in case data does need to be handled in the browser. We won't go into how Java-Script can be used for massive calculations within collaborative distributed frameworks, but an army of dwarfs can topple a giant. If this topic interests you, you can read more about it at https://www.igvita.com/2009/03/03/collaborative-map-reduce-in-the-browser/ and at http://dyn.com/blog/browsers-vs-servers-using-javascript-for-number-crunching-theories/.

d3.js can safely be called the most versatile JavaScript data visualization library available at the time of writing; it was developed by Mike Bostock as a successor to his Protovis library. Many JavaScript libraries are built on top of d3.js.

NVD3, C3.js, xCharts, and Dimple offer roughly the same thing: an abstraction layer on top of d3.js, which makes it easier to draw simple graphs. They mainly differ in the type of graphs they support and their default design. Feel free to visit their websites and find out for yourself:

- *NVD3*—http://nvd3.org/
- *C3.js*—http://c3js.org/
- *xCharts*—http://tenxer.github.io/xcharts/
- *Dimple*—http://dimplejs.org/

Many options exist. So why dc.js?

The main reason: compared to what it delivers, an interactive dashboard where clicking one graph will create filtered views on related graphs, dc.js is surprisingly easy to set up. It's so easy that you'll have a working example by the end of this chapter. As a data scientist, you already put in enough time on your actual analysis; easy-to-implement dashboards are a welcome gift.

To get an idea of what you're about to create, you can go to the following website, http://dc-js.github.io/dc.js/, and scroll down to the NASDAQ example, shown in figure 9.2.

Click around the dashboard and see the graphs react and interact when you select and deselect data points. Don't spend too long though; it's time to create this yourself.

As stated before, dc.js has two big prerequisites: d3.js and crossfilter.js. d3.js has a steep learning curve and there are several books on the topic worth reading if you're interested in full customization of your visualizations. But to work with dc.js, no knowledge of it is required, so we won't go into it in this book. Crossfilter.js is another matter; you'll need to have a little grasp of this MapReduce library to get dc.js up and running on your data. But because the concept of MapReduce itself isn't new, this will go smoothly.

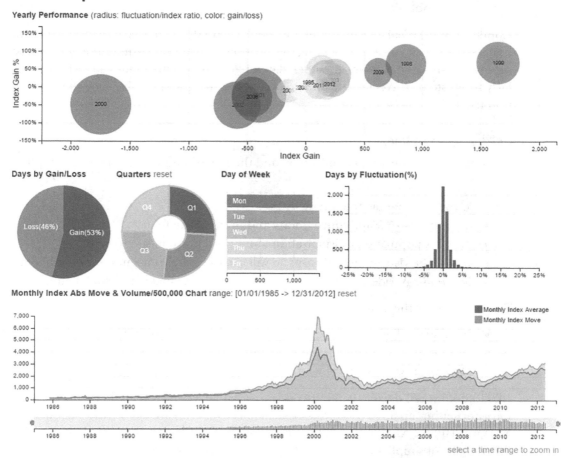

Figure 9.2 A dc.js interactive example on its official website

9.2   *Crossfilter, the JavaScript MapReduce library*

JavaScript isn't the greatest language for data crunching. But that didn't stop people, like the folks at Square, from developing MapReduce libraries for it. If you're dealing with data, every bit of speed gain helps. You don't want to send enormous loads of data over the internet or even your internal network though, for these reasons:

- Sending a bulk of data will tax the *network* to the point where it will bother other users.
- The *browser* is on the receiving end, and while loading in the data it will temporarily *freeze*. For small amounts of data this is unnoticeable, but when you start looking at 100,000 lines, it can become a visible lag. When you go over

1,000,000 lines, depending on the width of your data, your browser could give up on you.

Conclusion: it's a balance exercise. For the data you do send, there is a Crossfilter to handle it for you once it arrives in the browser. In our case study, the pharmacist requested the central server for stock data of 2015 for 29 medicines she was particularly interested in. We already took a look at the data, so let's dive into the application itself.

9.2.1   Setting up everything

It's time to build the actual application, and the ingredients of our small dc.js application are as follows:

- *JQuery*—To handle the interactivity
- *Crossfilter.js*—A MapReduce library and prerequisite to dc.js
- *d3.js*—A popular data visualization library and prerequisite to dc.js
- *dc.js*—The visualization library you will use to create your interactive dashboard
- *Bootstrap*—A widely used layout library you'll use to make it all look better

You'll write only three files:

- *index.html*—The HTML page that contains your application
- *application.js*—To hold all the JavaScript code you'll write
- *application.css*—For your own CSS

In addition, you'll need to run our code on an HTTP server. You could go through the effort of setting up a LAMP (Linux, Apache, MySQL, PHP), WAMP (Windows, Apache, MySQL, PHP), or XAMPP (Cross Environment, Apache, MySQL, PHP, Perl) server. But for the sake of simplicity we won't set up any of those servers here. Instead you can do it with a single Python command. Use your command-line tool (Linux shell or Windows CMD) and move to the folder containing your index.html (once it's there). You should have Python installed for other chapters of this book so the following command should launch a Python HTTP server on your localhost.

```
python -m SimpleHTTPServer
```

For Python 3.4

```
python -m http.server 8000
```

As you can see in figure 9.3, an HTTP server is started on localhost port 8000. In your browser this translates to "localhost:8000"; putting "0.0.0.0:8000" won't work.

Figure 9.3 Starting up a simple Python HTTP server

Make sure to have all the required files available in the same folder as your index.html. You can download them from the Manning website or from their creators' websites.

- *dc.css and dc.min.js*—https://dc-js.github.io/dc.js/
- *d3.v3.min.js*—http://d3js.org/
- *crossfilter.min.js*—http://square.github.io/crossfilter/

Now we know how to run the code we're about to create, so let's look at the index.html page, shown in the following listing.

Listing 9.1   An initial version of index.html

```html
<html>
<head>
    <title>Chapter 10.  Data Science Application</title>

    <link rel="stylesheet" href="https://maxcdn.bootstrapcdn.com/bootstrap
/3.3.0/css/bootstrap.min.css">
    <link rel="stylesheet" href="https://maxcdn.bootstrapcdn.com/bootstrap/
3.3.0/css/bootstrap-theme.min.css">

    <link rel="stylesheet" href="dc.css">
    <link rel="stylesheet" href="application.css">
</head>
<body>

    <main class='container'>
        <h1>Chapter 10:  Data Science Application</h1>
        <div class="row">
            <div class='col-lg-12'>
                <div id="inputtable" class="well well-sm"></div>
            </div>
        </div>
        <div class="row">
            <div class='col-lg-12'>
                <div id="filteredtable" class="well well-sm"></div>
            </div>
        </div>
    </main>
```

All CSS is loaded here.

Make sure to have dc.css downloaded from the Manning download page or from the dc website: https://dc-js.github.io/dc.js/. It must be present in the same folder as index.html file.

Main container incorporates everything visible to user.

```
<script src="https://code.jquery.com/jquery-1.9.1.min.js"></script>
<script src="https://maxcdn.bootstrapcdn.com/bootstrap/3.3.0/js
/bootstrap.min.js"></script>

<script src="crossfilter.min.js"></script>
<script src="d3.v3.min.js"></script>
<script src="dc.min.js"></script>
<script src="application.js"></script>
</body>
</html>
```

Make sure to have crossfilter.min.js, d3.v3.min.js, and dc.min.js downloaded from their websites or from the Manning website. Crossfilter: http://square.github.io/crossfilter/, d3.js: http://d3js.org/, dc.min.js: https://dc-js.github.io/dc.js/.

All Javascript is loaded here.

No surprises here. The header contains all the CSS libraries you'll use, so we'll load our JavaScript at the end of the HTML body. Using a JQuery onload handler, your application will be loaded when the rest of the page is ready. You start off with two table placeholders: one to show what your input data looks like, `<div id="input-table"></div>`, and the other one will be used with Crossfilter to show a filtered table, `<div id="filteredtable"></div>`. Several Bootstrap CSS classes were used, such as "*well*", "*container*", the Bootstrap grid system with "*row*" and "*col-xx-xx*", and so on. They make the whole thing look nicer but they aren't mandatory. More information on the Bootstrap CSS classes can be found on their website at http://getbootstrap.com/css/.

Now that you have your HTML set up, it's time to show your data onscreen. For this, turn your attention to the application.js file you created. First, we wrap the entire code "to be" in a JQuery onload handler.

```
$(function() {
    //All future code will end up in this wrapper
})
```

Now we're certain our application will be loaded only when all else is ready. This is important because we'll use JQuery selectors to manipulate the HTML. It's time to load in data.

```
d3.csv('medicines.csv',function(data) {
    main(data)
});
```

You don't have a REST service ready and waiting for you, so for the example you'll draw the data from a .csv file. This file is available for download on Manning's website. d3.js offers an easy function for that. After loading in the data you hand it over to your main application function in the d3.csv callback function.

Apart from the main function you have a `CreateTable` function, which you will use to...you guessed it...create your tables, as shown in the following listing.

Listing 9.2 The `CreateTable` function

```
var tableTemplate = $([
    "<table class='table table-hover table-condensed table-striped'>",
    "  <caption></caption>",
    "  <thead><tr/></thead>",
    "  <tbody></tbody>",
    "</table>"
].join('\n'));

CreateTable = function(data,variablesInTable,title){
    var table = tableTemplate.clone();
    var ths = variablesInTable.map(function(v) { return $("<th>").text(v)
});
    $('caption', table).text(title);
    $('thead tr', table).append(ths);
    data.forEach(function(row) {
        var tr = $("<tr>").appendTo($('tbody', table));
        variablesInTable.forEach(function(varName) {
            var val = row, keys = varName.split('.');
            keys.forEach(function(key) { val = val[key] });
            tr.append($("<td>").text(val));
        });
    });
    return table;
}
```

`CreateTable()` requires three arguments:

- `data`—The data it needs to put into a table.
- `variablesInTable`—What variables it needs to show.
- `Title`—The title of the table. It's always nice to know what you're looking at.

`CreateTable()` uses a predefined variable, `tableTemplate`, that contains our overall table layout. `CreateTable()` can then add rows of data to this template.

Now that you have your utilities, let's get to the `main` function of the application, as shown in the following listing.

Listing 9.3 JavaScript `main` function

```
main = function(inputdata){
                                              ◄─── Our data: normally this is fetched
                                                   from a server but in this case we
    var medicineData = inputdata ;            ◄─── read it from a local .csv file

    var dateFormat = d3.time.format("%d/%m/%Y");    ◄─── Convert date to correct
    medicineData.forEach(function (d) {                  format so Crossfilter will
        d.Day = dateFormat.parse(d.Date);                recognize date variable
    })
    var variablesInTable =
    ['MedName','StockIn','StockOut','Stock','Date','LightSen']    ◄─── Only show a
        var sample = medicineData.slice(0,5);                         sample of data
        var inputTable = $("#inputtable");    ◄─── Create table
```

Put the variables we'll show in the table in an array so we can loop through them when creating table code

```
inputTable
        .empty()
        .append(CreateTable(sample,variablesInTable,"The input table"));
}
```

You start off by showing your data on the screen, but preferably not all of it; only the first five entries will do, as shown in figure 9.4. You have a date variable in your data and you want to make sure Crossfilter will recognize it as such later on, so you first parse it and create a new variable called Day. You show the original, Date, to appear in the table for now, but later on you'll use Day for all your calculations.

The input table

MedName	StockIn	StockOut	Stock	Date	LightSen
Acupan 30 mg	150	-7	143	1/01/2015	No
Acupan 30 mg	5	-6	142	2/01/2015	No
Acupan 30 mg	15	-9	148	3/01/2015	No
Acupan 30 mg	0	-11	137	4/01/2015	No
Acupan 30 mg	10	-8	139	5/01/2015	No

Figure 9.4   Input medicine table shown in browser: first five lines

This is what you end up with: the same thing you saw in Excel before. Now that you know the basics are working, you'll introduce Crossfilter into the equation.

9.2.2   *Unleashing Crossfilter to filter the medicine data set*

Now let's go into Crossfilter to use filtering and MapReduce. Henceforth you can put all the upcoming code after the code of section 9.2.1 within the main() function. The first thing you'll need to do is declare a Crossfilter instance and initiate it with your data.

```
CrossfilterInstance = crossfilter(medicineData);
```

From here you can get to work. On this instance you can register dimensions, which are the columns of your table. Currently Crossfilter is limited to 32 dimensions. If you're handling data wider than 32 dimensions, you should consider narrowing it down before sending it to the browser. Let's create our first dimension, the medicine name dimension:

```
var medNameDim = CrossfilterInstance.dimension(function(d) {return
    d.MedName;});
```

Your first dimension is the name of the medicines, and you can already use this to filter your data set and show the filtered data using our `CreateTable()` function.

```
var dataFiltered= medNameDim.filter('Grazax 75 000 SQ-T')
var filteredTable = $('#filteredtable');
filteredTable
.empty().append(CreateTable(dataFiltered.top(5),variablesInTable,'Our
First Filtered Table'));
```

You show only the top five observations (figure 9.5); you have 365 because you have the results from a single medicine for an entire year.

Our First Filtered Table

MedName	StockIn	StockOut	Stock	Date	LightSen
Grazax 75 000 SQ-T	15	0	205	31/08/2015	Yes
Grazax 75 000 SQ-T	0	-4	62	30/12/2015	Yes
Grazax 75 000 SQ-T	10	-15	66	29/12/2015	Yes
Grazax 75 000 SQ-T	15	0	71	28/12/2015	Yes
Grazax 75 000 SQ-T	10	-4	56	27/12/2015	Yes

Figure 9.5   Data filtered on medicine name Grazax 75 000 SQ-T

This table doesn't look sorted but it is. The `top()` function sorted it on medicine name. Because you only have a single medicine selected it doesn't matter. Sorting on date is easy enough using your new `Day` variable. Let's register another dimension, the date dimension:

```
var DateDim = CrossfilterInstance.dimension(
function(d) {return d.Day;});
```

Now we can sort on date instead of medicine name:

```
filteredTable
        .empty()
        .append(CreateTable(DateDim.bottom(5),variablesInTable,'Our
First Filtered Table'));
```

The result is a bit more appealing, as shown in figure 9.6.

This table gives you a window view of your data but it doesn't summarize it for you yet. This is where the Crossfilter MapReduce capabilities come in. Let's say you would like to know how many observations you have per medicine. Logic dictates that you

| Our First Filtered Table | | | | | |
MedName	StockIn	StockOut	Stock	Date	LightSen
Grazax 75 000 SQ-T	65	-12	53	1/01/2015	Yes
Grazax 75 000 SQ-T	15	-11	57	2/01/2015	Yes
Grazax 75 000 SQ-T	5	-9	53	3/01/2015	Yes
Grazax 75 000 SQ-T	5	-4	54	4/01/2015	Yes
Grazax 75 000 SQ-T	0	-14	40	5/01/2015	Yes

Figure 9.6   Data filtered on medicine name Grazax 75 000 SQ-T and sorted by day

should end up with the same number for every medicine: 365, or 1 observation per day in 2015.

```
var countPerMed = medNameDim.group().reduceCount();
variablesInTable = ["key","value"]
filteredTable
        .empty()
  .append(CreateTable(countPerMed.top(Infinity),
variablesInTable,'Reduced Table'));
```

Crossfilter comes with two MapReduce functions: `reduceCount()` and `reduceSum()`. If you want to do anything apart from counting and summing, you need to write reduce functions for it. The `countPerMed` variable now contains the data grouped by the medicine dimension and a line count for each medicine in the form of a key and a value. To create the table you need to address the variable `key` instead of `medName` and `value` for the count (figure 9.7).

| Reduced Table | |
key	value
Adoport 1 mg	365
Atenolol EG 100 mg	365
Ceftriaxone Actavis 1 g	365
Cefuroxim Mylan 500 mg	365
Certican 0.25 mg	365

Figure 9.7   MapReduced table with the medicine as the group and a count of data lines as the value

By specifying `.top(Infinity)` you ask to show all 29 medicines onscreen, but for the sake of saving paper figure 9.7 shows only the first five results. Okay, you can rest easy; the data contains 365 lines per medicine. Notice how Crossfilter ignored the filter on "Grazax". If a dimension is used for grouping, the filter doesn't apply to it. Only filters on other dimensions can narrow down the results.

What about more interesting calculations that don't come bundled with Crossfilter, such as an average, for instance? You can still do that but you'd need to write three functions and feed them to a `.reduce()` method. Let's say you want to know the average stock per medicine. As previously mentioned, almost all of the MapReduce logic needs to be written by you. An average is nothing more than the division of sum by count, so you will require both; how do you go about this? Apart from the `reduceCount()` and `reduceSum()` functions, Crossfilter has the more general `reduce()` function. This function takes three arguments:

- *The reduceAdd() function*—A function that describes what happens when an extra observation is added.
- *The reduceRemove() function*—A function that describes what needs to happen when an observation disappears (for instance, because a filter is applied).
- *The reduceInit() function*—This one sets the initial values for everything that's calculated. For a sum and count the most logical starting point is 0.

Let's look at the individual reduce functions you'll require before trying to call the Crossfilter `.reduce()` method, which takes these three components as arguments. A custom reduce function requires three components: an initiation, an add function, and a remove function. The initial reduce function will set starting values of the p object:

```
var reduceInitAvg = function(p,v){
    return {count: 0, stockSum : 0, stockAvg:0};
}
```

As you can see, the reduce functions themselves take two arguments. These are automatically fed to them by the Crossfilter `.reduce()` method:

- p is an object that contains the combination situation so far; it persists over all observations. This variable keeps track of the sum and count for you and thus represents your goal, your end result.
- v represents a record of the input data and has all its variables available to you. Contrary to p, it doesn't persist but is replaced by a new line of data every time the function is called. The `reduceInit()` is called only once, but `reduceAdd()` is called every time a record is added and `reduceRemove()` every time a line of data is removed.
- The `reduceInit()` function, here called `reduceInitAvg()` because you're going to calculate an average, basically initializes the p object by defining its components (count, sum, and average) and setting their initial values. Let's look at `reduceAddAvg()`:

```
var reduceAddAvg = function(p,v){
    p.count += 1;
    p.stockSum  = p.stockSum  + Number(v.Stock);
    p.stockAvg = Math.round(p.stockSum  / p.count);
    return p;
}
```

reduceAddAvg() takes the same p and v arguments but now you actually use v; you don't need your data to set the initial values of p in this case, although you can if you want to. Your Stock is summed up for every record you add, and then the average is calculated based on the accumulated sum and record count:

```
var reduceRemoveAvg = function(p,v){
    p.count -= 1;
    p.stockSum  = p.stockSum  -  Number(v.Stock);
    p.stockAvg = Math.round(p.stockSum  / p.count);
    return p;
}
```

The reduceRemoveAvg() function looks similar but does the opposite: when a record is removed, the count and sum are lowered. The average always calculates the same way, so there's no need to change that formula.

The moment of truth: you apply this homebrewed MapReduce function to the data set:

```
        dataFiltered = medNameDim.group().reduce(reduceAddAvg,
reduceRemoveAvg,reduceInitAvg)

        variablesInTable = ["key","value.stockAvg"]
        filteredTable
            .empty()
    .append(CreateTable(dataFiltered.top(Infinity),
variablesInTable,'Reduced Table'));
```

Business as usual: draw result table.

reduce() takes the 3 functions (reduceInitAvg(), reduceAddAvg(), and reduceRemoveAvg()) as input arguments.

Notice how the name of your output variable has changed from value to value .stockAvg. Because you defined the reduce functions yourself, you can output many variables if you want to. Therefore, value has changed into an object containing all the variables you calculated; stockSum and count are also in there.

The results speak for themselves, as shown in figure 9.8. It seems we've borrowed Cimalgex from other hospitals, going into an average negative stock.

This is all the Crossfilter you need to know to work with dc.js, so let's move on and bring out those interactive graphs.

Reduced Table	
key	value.stockAvg
Adoport 1 mg	36
Atenolol EG 100 mg	49
Ceftriaxone Actavis 1 g	207
Cefuroxim Mylan 500 mg	118
Certican 0.25 mg	158
Cimalgex 8 mg	-24

Figure 9.8   MapReduced table with average stock per medicine

9.3 Creating an interactive dashboard with dc.js

Now that you know the basics of Crossfilter, it's time to take the final step: building the dashboard. Let's kick off by making a spot for your graphs in the index.html page. The new body looks like the following listing. You'll notice it looks similar to our initial setup apart from the added graph placeholder <div> tags and the reset button <button> tag.

Listing 9.4 A revised index.html with space for graphs generated by dc.js

```
Layout:
        Title
        | input table                  | (row 1)
        | filtered table               | (row 2)
        [ reset button ]
        | stock-over-time chart | stock-per-medicine chart | (row 3)
        | light-sensitive chart |                          | (row 4)
            (column 1)          (column 2)


<body>
    <main class='container'>

        <h1>Chapter 10:  Data Science Application</h1>
        <div class="row">
            <div class='col-lg-12'>
                <div id="inputtable" class="well well-sm">
</div>                                            ⊲──        This is a placeholder
                                                            <div> for input data
            </div>                                          table inserted later.
        </div>
        <div class="row">
            <div class='col-lg-12'>
                <div id="filteredtable" class="well well-sm">
</div>                                            ⊲──        This is a placeholder
                                                            <div> for filtered
            </div>                                          table inserted later.
        </div>
```

This is new: reset button. ⟿

```
        <button class="btn btn-success">Reset Filters</button>
        <div class="row">
            <div class="col-lg-6">
                <div id="StockOverTime" class="well well-sm"></div>  ⊲──
```

This is new: time chart placeholder.

This is new: light sensitivity pie-chart placeholder. ⟿

```
                <div id="LightSensitiveStock" class="well well-sm"></div>

            </div>
            <div class="col-lg-6">
                <div id="StockPerMedicine" class="well well-sm"></div>  ⊲──
            </div>
        </div>
    </main>
```

This is new: stock per medicine bar-chart placeholder.

```
     ┌──► <script src="https://code.jquery.com/jquery-1.9.1.min.js"></script>
     │    <script src="https://maxcdn.bootstrapcdn.com/bootstrap
/3.3.0/js/bootstrap.min.js"></script>

          <script src="crossfilter.min.js"></script>        │ Crossfilter, d3, and dc libraries
          <script src="d3.v3.min.js"></script>              │ can be downloaded from their
          <script src="dc.min.js"></script>                 │ respective websites.

          <script src="application.js"></script>   ◄──┐ Our own application
     </body>                                            │ JavaScript code.
```

Standard practice. JS libraries are last to speed page load.
    jQuery: vital HTML-JavaScript interaction
    Bootstrap: simplified CSS and layout from folks at Twitter
    Crossfilter: our JavaScript MapReduce library of choice
    d3:   the d3 script, necessary to run dc.js
    DC: our visualization library
    application: our data science application; here we store all the logic
   *.min.js denotes minified JavaScript for our 3rd party libraries

We've got Bootstrap formatting going on, but the most important elements are the three <div> tags with IDs and the button. What you want to build is a representation of the total stock over time, <div id="StockOverTime"></div>, with the possibility of filtering on medicines, <div id="StockPerMedicine"></div>, and whether they're light-sensitive or not, <div id="LightSensitiveStock"></div>. You also want a button to reset all the filters, <button class="btn btn-success">Reset Filters</button>. This reset button element isn't required, but is useful.

Now turn your attention back to application.js. In here you can add all the upcoming code in your main() function as before. There is, however, one exception to the rule: dc.renderAll(); is dc's command to draw the graphs. You need to place this render command only once, at the bottom of your main() function. The first graph you need is the "total stock over time," as shown in the following listing. You already have the time dimension declared, so all you need is to sum your stock by the time dimension.

Listing 9.5   Code to generate "total stock over time" graph

```
         var SummatedStockPerDay =                            │ Stock over
     DateDim.group().reduceSum(function(d){return d.Stock;})  ◄──┘ time data

         var minDate = DateDim.bottom(1)[0].Day;
         var maxDate = DateDim.top(1)[0].Day;                 │ Line
         var StockOverTimeLineChart = dc.lineChart("#StockOverTime"); │ chart

         StockOverTimeLineChart
 Deliveries      .width(null) // null means size to fit container
 per day         .height(400)
 graph           .dimension(DateDim)
                 .group(SummatedStockPerDay)
```

```
                        .x(d3.time.scale().domain([minDate,maxDate]))
  Deliveries            .xAxisLabel("Year 2015")
  per day               .yAxisLabel("Stock")
  graph                 .margins({left: 60, right: 50, top: 50, bottom: 50})

  dc.renderAll();                    Render all
                                     graphs
```

Look at all that's happening here. First you need to calculate the range of your x-axis so dc.js will know where to start and end the line chart. Then the line chart is initialized and configured. The least self-explanatory methods here are .group() and .dimension(). .group() takes the time dimension and represents the x-axis. .dimension() is its counterpart, representing the y-axis and taking your summated data as input. Figure 9.9 looks like a boring line chart, but looks can be deceiving.

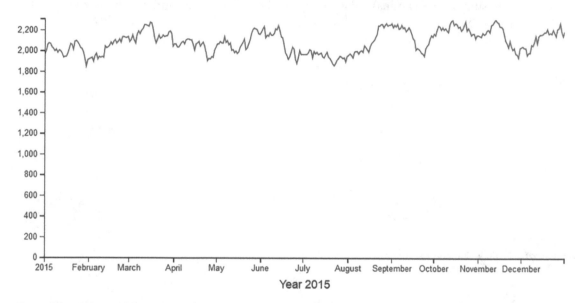

Figure 9.9   dc.js graph: sum of medicine stock over the year 2015

Things change drastically once you introduce a second element, so let's create a row chart that represents the average stock per medicine, as shown in the next listing.

Listing 9.6   Code to generate "average stock per medicine" graph

```
var AverageStockPerMedicineRowChart = dc.rowChart("#StockPerMedicine");
var AvgStockMedicine = medNameDim.group().reduce(reduceAddAvg,
reduceRemoveAvg,reduceInitAvg);

AverageStockPerMedicineRowChart
          .width(null)              Null means "size        Average stock
          .height(1200)             to fit container"        per medicine
                                                              row chart
```

```
.dimension(medNameDim)
.group(AvgStockMedicine)
.margins({top: 20, left: 10, right: 10, bottom: 20})
.valueAccessor(function (p) {return p.value.stockAvg;});
```

This should be familiar because it's a graph representation of the table you created earlier. One big point of interest: because you used a custom-defined `reduce()` function this time, dc.js doesn't know what data to represent. With the `.valueAccessor()` method you can specify `p.value.stockAvg` as the value of your choice. The dc.js row chart's label's font color is gray; this makes your row chart somewhat hard to read. You can remedy this by overwriting its CSS in your application.css file:

```
.dc-chart g.row text {fill: black;}
```

One simple line can make the difference between a clear and an obscure graph (figure 9.10).

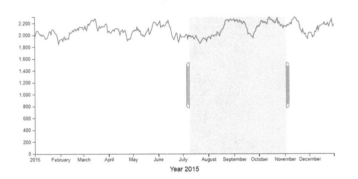

Figure 9.10   dc.js line chart and row chart interaction

Now when you select an area on the line chart, the row chart is automatically adapted to represent the data for the correct time period. Inversely, you can select one or multiple medicines on the row chart, causing the line chart to adjust accordingly. Finally, let's add the light-sensitivity dimension so the pharmacist can distinguish between stock for light-sensitive medicines and non-light-sensitive ones, as shown in the following listing.

Listing 9.7   Adding the light-sensitivity dimension

```
        var lightSenDim = CrossfilterInstance.dimension(
function(d){return d.LightSen;});
        var SummatedStockLight =  lightSenDim.group().reduceSum(
function(d) {return d.Stock;});

        var LightSensitiveStockPieChart = dc.pieChart("#LightSensitiveStock");
```

```
LightSensitiveStockPieChart
        .width(null) // null means size to fit container
        .height(300)
        .dimension(lightSenDim)
        .radius(90)
        .group(SummatedStockLight)
```

We hadn't introduced the light dimension yet, so you need to register it onto your Crossfilter instance first. You can also add a reset button, which causes all filters to reset, as shown in the following listing.

Listing 9.8 The dashboard reset filters button

When an element with class btn-success is clicked (our reset button), resetFilters() is called.

```
resetFilters = function(){
    StockOverTimeLineChart.filterAll();
    LightSensitiveStockPieChart.filterAll();
    AverageStockPerMedicineRowChart.filterAll();
    dc.redrawAll();
}
$('.btn-success').click(resetFilters);
```

resetFilters() function will reset our dc.js data and redraw graphs.

The `.filterAll()` method removes all filters on a specific dimension; `dc.redraw-All()` then manually triggers all dc charts to redraw.

The final result is an interactive dashboard (figure 9.11), ready to be used by our pharmacist to gain insight into her stock's behavior.

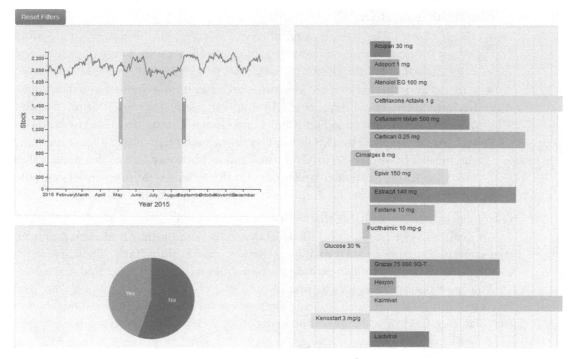

Figure 9.11 dc.js fully interactive dashboard on medicines and their stock within the hospital pharmacy

9.4   *Dashboard development tools*

We already have our glorious dashboard, but we want to end this chapter with a short (and far from exhaustive) overview of the alternative software choices when it comes to presenting your numbers in an appealing way.

You can go with proven and true software packages of renowned developers such as Tableau, MicroStrategy, Qlik, SAP, IBM, SAS, Microsoft, Spotfire, and so on. These companies all offer dashboard tools worth investigating. If you're working in a big company, chances are good you have at least one of those paid tools at your disposal. Developers can also offer free public versions with limited functionality. Definitely check out Tableau if you haven't already at http://www.tableausoftware.com/public/download.

Other companies will at least give you a trial version. In the end you have to pay for the full version of any of these packages, and it might be worth it, especially for a bigger company that can afford it.

This book's main focus is on free tools, however. When looking at free data visualization tools, you quickly end up in the HTML world, which proliferates with free JavaScript libraries to plot any data you want. The landscape is enormous:

- *HighCharts*—One of the most mature browser-based graphing libraries. The free license applies only to noncommercial pursuits. If you want to use it in a commercial context, prices range anywhere from $90 to $4000. See http://shop .highsoft.com/highcharts.html.
- *Chartkick*—A JavaScript charting library for Ruby on Rails fans. See http://ank-ane.github.io/chartkick/.
- *Google Charts*—The free charting library of Google. As with many Google products, it is free to use, even commercially, and offers a wide range of graphs. See https://developers.google.com/chart/.
- *d3.js*—This is an odd one out because it isn't a graphing library but a data visualization library. The difference might sound subtle but the implications are not. Whereas libraries such as HighCharts and Google Charts are meant to draw certain predefined charts, d3.js doesn't lay down such restrictions. d3.js is currently the most versatile JavaScript data visualization library available. You need only a quick peek at the interactive examples on the official website to understand the difference from a regular graph-building library. See http://d3js.org/.

Of course, others are available that we haven't mentioned.

You can also get visualization libraries that only come with a trial period and no free community edition, such as Wijmo, Kendo, and FusionCharts. They are worth looking into because they also provide support and guarantee regular updates.

You have options. But why or when would you even consider building your own interface with HTML5 instead of using alternatives such as SAP's BusinessObjects, SAS JMP, Tableau, Clickview, or one of the many others? Here are a few reasons:

- *No budget*—When you work in a startup or other small company, the licensing costs accompanying this kind of software can be high.

- *High accessibility*—The data science application is meant to release results to any kind of user, especially people who might only have a browser at their disposal—your own customers, for instance. Data visualization in HTML5 runs fluently on mobile.

- *Big pools of talent out there*—Although there aren't that many Tableau developers, scads of people have web-development skills. When planning a project, it's important to take into account whether you can staff it.

- *Quick release*—Going through the entire IT cycle might take too long at your company, and you want people to enjoy your analysis quickly. Once your interface is available and being used, IT can take all the time they want to industrialize the product.

- *Prototyping*—The better you can show IT its purpose and what it should be capable of, the easier it is for them to build or buy a sustainable application that does what you want it to do.

- *Customizability*—Although the established software packages are great at what they do, an application can never be as customized as when you create it yourself.

And why wouldn't you do this?

- *Company policy*—This is the biggest one: it's not allowed. Large companies have IT backup teams that allow only a certain number of tools to be used so they can keep their supporting role under control.

- *You have an experienced team of reporters at your disposal*—You'd be doing their job, and they might come after you with pitchforks.

- *Your tool does allow enough customization to suit your taste*—Several of the bigger platforms are browser interfaces with JavaScript running under the hood. Tableau, BusinessObjects Webi, SAS Visual Analytics, and so on all have HTML interfaces; their tolerance to customization might grow over time.

The front end of any application can win the hearts of the crowd. All the hard work you put into data preparation and the fancy analytics you applied is only worth as much as you can convey to those who use it. Now you're on the right track to achieve this. On this positive note we'll conclude this chapter.

9.5 Summary

- This chapter focused on the last part of the data science process, and our goal was to build a data science application where the end user is provided with an interactive dashboard. After going through all the steps of the data science process, we're presented with clean, often compacted or information dense, data. This way we can query less data and get the insights we want.

- In our example, the pharmacy stock data is considered thoroughly cleaned and prepared and this should always be the case by the time the information reaches the end user.

- JavaScript-based dashboards are perfect for quickly granting access to your data science results because they only require the user to have a web browser. Alternatives exist, such as Qlik (chapter 5).

- Crossfilter is a MapReduce library, one of many JavaScript MapReduce libraries, but it has proven its stability and is being developed and used by Square, a company that does monetary transactions. Applying MapReduce is effective, even on a single node and in a browser; it increases the calculation speed.

- dc.js is a chart library build on top of d3.js and Crossfilter that allows for quick browser dashboard building.

- We explored the data set of a hospital pharmacy and built an interactive dashboard for pharmacists. The strength of a dashboard is its *self-service* nature: they don't always need a reporter or data scientist to bring them the insights they crave.

- Data visualization alternatives are available, and it's worth taking the time to find the one that suits your needs best.

- There are multiple reasons why you'd create your own custom reports instead of opting for the (often more expensive) company tools out there:
  - *No budget*—Startups can't always afford every tool
  - *High accessibility*—Everyone has a browser
  - *Available talent*—(Comparatively) easy access to JavaScript developers
  - *Quick release*—IT cycles can take a while
  - *Prototyping*—A prototype application can provide and leave time for IT to build the production version
  - *Customizability*—Sometimes you just want it exactly as your dreams picture it.

- Of course there are reasons against developing your own application:
  - *Company policy*—Application proliferation isn't a good thing and the company might want to prevent this by restricting local development.
  - *Mature reporting team*—If you have a good reporting department, why would you still bother?
  - *Customization is satisfactory*—Not everyone wants the shiny stuff; basic can be enough.

Congratulations! You've made it to the end of this book and the true beginning of your career as a data scientist. We hope you had ample fun reading and working your way through the examples and case studies. Now that you have basic insight into the world of data science, it's up to you to choose a path. The story continues, and we all wish you great success in your quest of becoming the greatest data scientist who has ever lived! May we meet again someday. ;)

appendix A
Setting up Elasticsearch

In this appendix, we'll cover installing and setting up the Elasticsearch database used in Chapters 6 and 7. Instructions for both Linux and Windows installations are included. Note that if you get into trouble or want further information on Elasticsearch, it has pretty decent documentation you can find located at https://www.elastic.co/guide/en/elasticsearch/reference/1.4/setup.html.

> **NOTE** Elasticsearch is dependent on Java, so we'll cover how to install that as well.

A.1 Linux installation

First check to see if you have Java already installed on your machine.

1 You can check your Java version in a console window with `java -version`. If Java is installed, you'll see a response like the one in figure A.1. You'll need at least Java 7 to run the version of Elasticsearch we use in this book (1.4). Note: Elasticsearch had moved on to version 2 by the time this book was released, but while code might change slightly, the core principles remain the same.

Figure A.1   Checking the Java version in Linux. Elasticsearch requires Java 7 or higher.

2 If Java isn't installed or you don't have a high enough version, Elasticsearch recommends the Oracle version of Java. Use the following console commands to install it.

```
sudo add-apt-repository ppa:webupd8team/java
sudo apt-get install oracle-java7-installer
```

Now you can install Elasticsearch:

1 Add the Elasticsearch 1.4 repo, which is the latest one at the time of writing, to your repo list and then install it with the following commands.

```
sudo add-apt-repository "deb http://packages.Elasticsearch.org/
➥                             Elasticsearch/1.4/debian stable main"
sudo apt-get update && sudo apt-get install Elasticsearch
```

2 To make sure Elasticsearch will start on reboot, run the following command.

```
sudo update-rc.d Elasticsearch defaults 95 10
```

3 Turn on Elasticsearch. See figure A.2.

```
sudo /etc/init.d/Elasticsearch start
```

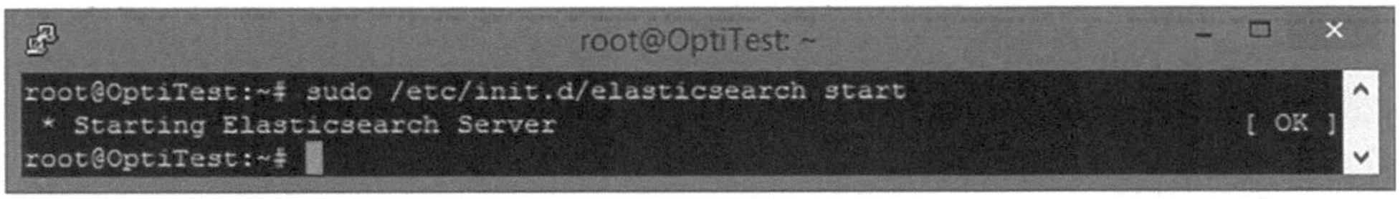
root@OptiTest: ~
root@OptiTest:~# sudo /etc/init.d/elasticsearch start
* Starting Elasticsearch Server
[OK]
root@OptiTest:~#

Figure A.2 Starting Elasticsearch on Linux

If Linux is your local computer, open a browser and go to *localhost:9200*. 9200 is the default port for the Elasticsearch API. See figure A.3.

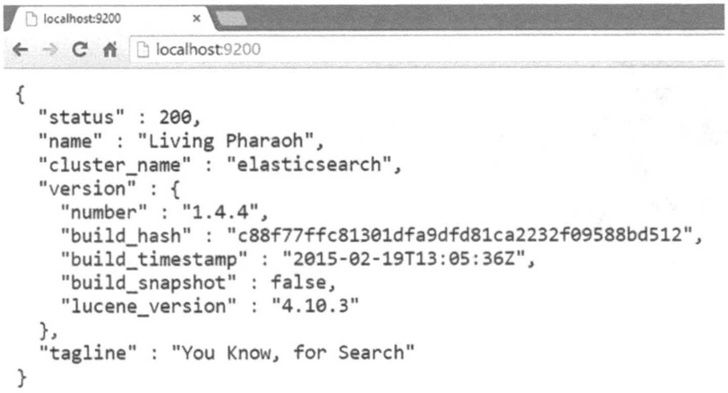
localhost:9200
localhost:9200
{
"status" : 200,
"name" : "Living Pharaoh",
"cluster_name" : "elasticsearch",
"version" : {
"number" : "1.4.4",
"build_hash" : "c88f77ffc81301dfa9dfd81ca2232f09588bd512",
"build_timestamp" : "2015-02-19T13:05:36Z",
"build_snapshot" : false,
"lucene_version" : "4.10.3"
},
"tagline" : "You Know, for Search"
}

Figure A.3 The Elasticsearch welcome screen on localhost

The Elasticsearch welcome screen should greet you. Notice your database even has a name. The name is picked from the pool of Marvel characters and changes every time you reboot your database. In production, having an inconsistent and non-unique name such as this can be problematic. The instance you started is a single node of what could be part of a huge distributed cluster. If all of these nodes change names on reboot, it becomes nearly impossible to track them with logs in case of trouble. Elasticsearch takes pride in the fact it has little need for configuration to get you started and is distributed by nature. While this is most certainly true, things such as this random name prove that deploying an actual multi-node setup will require you to think twice about certain default settings. Luckily Elasticsearch has adequate documentation on almost everything, including deployment (http://www.Elasticsearch.org/guide/en/Elasticsearch/guide/current/deploy.html). Multi-node Elasticsearch deployment isn't in the scope of this chapter but it's good to keep in mind.

A.2 *Windows installation*

InWindows, Elasticsearch also requires at least Java 7—the JRE and the JDK—to be installed and for the JAVA_HOME variable to be pointing at the Java folder.

1 Download the Windows installers for Java from http://www.oracle.com/technetwork/java/javase/downloads/index.html and run them.

2 After installation make sure your JAVA_HOME Windows environment variable points to where you installed the Java Development Kit. You can find your environment variables in System Control Panel > Advanced System Settings. See figure A.4.

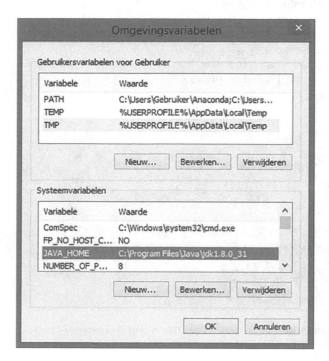

Figure A.4 The JAVA_HOME variable set to the Java install folder

Attempting an install before you have an adequate Java version will result in an error. See figure A.5.

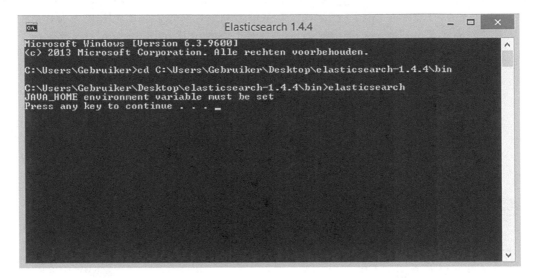

Figure A.5   The Elasticsearch install fails when JAVA_HOME is not set correctly.

Installing on a PC with limited rights

Sometimes you want to try a piece of software but you aren't free to install your own programs. If that's the case, don't despair: portable JDKs are out there. When you find one of those you can temporarily set your JAVA_HOME variable to the path of the portable JDK and start Elasticsearch this way. You don't even need to install Elasticsearch if you're only checking it out. See figure A.6.

Figure A.6   Starting Elasticsearch without an installation. This is only recommended for testing purposes on a computer where you have limited rights.

Now that you have Java installed and set up, you can install Elasticsearch.

1 Download the Elasticsearch zip package manually from http://www.Elastic-search.org/download/. Unpack it anywhere on your computer. This folder will now become your self-contained database. If you have an SSD drive, consider giving it a place there, because it significantly increases the speed of Elasticsearch.

2 If you already have a Windows command window open, don't use it for the installation; open a fresh one instead. The environment variables in the open window aren't up to date anymore. Change the directory to your Elasticsearch /bin folder and install using the `service install` command. See figure A.7.

Figure A.7 An Elasticsearch Windows 64-bit installation

3 The database should now be ready to start. Use the `service start` command. See figure A.8.

Figure A.8 Elasticsearch starts up a node on Windows.

If you want to stop the server, issue the `service stop` command. Open your browser of choice and put *localhost:9200* in the address bar. If the Elasticsearch welcome screen appears (figure A.9), you've successfully installed Elasticsearch.

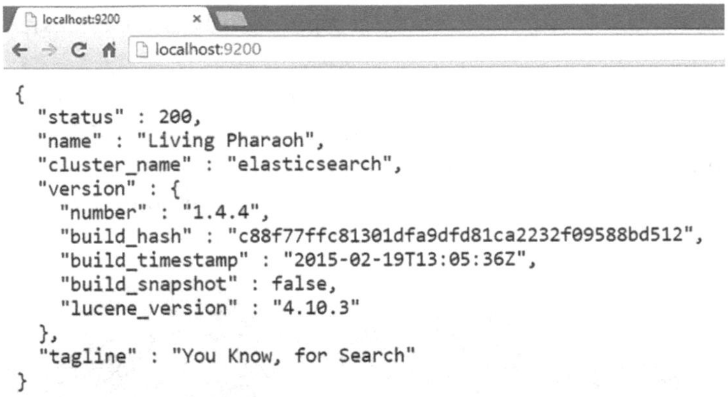
localhost:9200
localhost:9200
{
"status" : 200,
"name" : "Living Pharaoh",
"cluster_name" : "elasticsearch",
"version" : {
"number" : "1.4.4",
"build_hash" : "c88f77ffc81301dfa9dfd81ca2232f09588bd512",
"build_timestamp" : "2015-02-19T13:05:36Z",
"build_snapshot" : false,
"lucene_version" : "4.10.3"
},
"tagline" : "You Know, for Search"
}

Figure A.9 The Elasticsearch welcome screen on localhost

appendix B
Setting up Neo4j

In this appendix, we'll cover installing and setting up the Neo4j community edition database used in Chapter 7. Instructions for both Linux and Windows installations are included.

B.1 Linux installation

To install Neo4j community edition on Linux, use your command line as instructed here: http://debian.neo4j.org/?_ga=1.84149595.332593114.1442594242.

Neo Technology provides this Debian repository to make it easy to install Neo4j. It includes three repositories:

- *Stable*—All Neo4j releases, except as noted below. You should choose this by default.
- *Testing*—Pre-release versions (milestones and release candidates).
- *Oldstable*—No longer actively used, this repository contains patch releases for old minor versions. If you can't find what you need in Stable, then look here.

To use the new Stable packages, you need to run the commands below as root (note that we use sudo below):

```
sudo -s
wget -O - https://debian.neo4j.org/neotechnology.gpg.key| apt-key add - #
    Import our signing key
echo 'deb http://debian.neo4j.org/repo stable/' > /etc/apt/sources.list.d/
    neo4j.list # Create an Apt sources.list file
aptitude update -y # Find out about the files in our repository
aptitude install neo4j -y # Install Neo4j, community edition
```

You could replace Stable with Testing if you want a newer (but unsupported) build of Neo4j. If you'd like a different edition, you can run:

```
apt-get install neo4j-advanced
```

or

```
apt-get install neo4j-enterprise
```

B.2   *Windows installation*

To install the Neo4j community edition on Windows:

1   Go to http://neo4j.com/download/ and download the community edition. The following screen will appear.

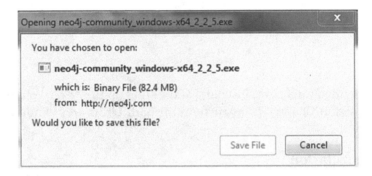

2   Save this file and run it.
3   After installation, you'll get a new pop up that gives you the option to choose the default database location or alternatively browse to find another location to use as the database location.
4   After making your choice, press Start and you're ready to go.

In a few seconds, the database will be ready to use. If you want to stop the server you can just press the Stop button.

5   Open your browser of choice and put *localhost:7474* in the address bar.

 You have arrived at the Neo4j browser.

6   When the database access asks for authentication, use the username and password *"neo4j"*, then press Connect.

 In the following window you can set your own password.

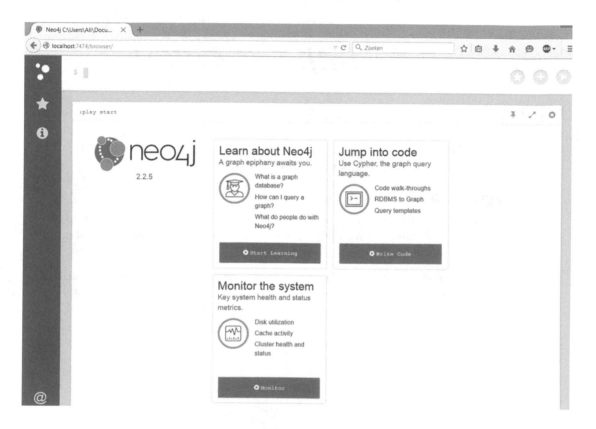

Now you can input your Cypher queries and consult your nodes, relationships, and results.

appendix C
Installing MySQL server

In this appendix, we'll cover installing and setting up the MySQL database. Instructions for Windows and Linux installations are included.

C.1   *Windows installation*

The most convenient and recommended method is to download MySQL installer (for Windows) and let it set up all of the MySQL components on your system. The following steps explain how to do it:

1. Download MySQL Installer from http://dev.mysql.com/downloads/installer/ and open it. Please notice that, unlike the standard MySQL installer, the smaller "web-group" version does automatically include any MySQL components, but will only download the ones you choose to install. Feel free to pick either installer. See figure C.1.

Figure C.1   Download options of MySQL installers for Windows

2 Select the suitable Setup Type you prefer. The option Developer Default will install MySQL server and other MySQL components related to MySQL advancement, together with supportive functions such as MySQL Workbench. You can also choose Custom Setup if you want to select the MySQL items that will be installed on your system. And you can always have different versions of MySQL operate on a single system, if you wish. The MySQL notifier is useful for monitoring the running instances, stopping them, and restarting them. You can also add this later using the MySQL installer.

3 Then the MySQL installation wizard's instructions will guide you through the setup process. It's mostly accepting what's to come. A development machine will do as the server configuration type. Make sure to set a MySQL root password and don't forget what it is, because you need it later. You can run it as a Windows service; that way, you don't need to launch it manually.

4 The installation completes. If you opted for a full install, by default MySQL server, MySQL workbench, and MySQL notifier will start automatically at computer startup. MySQL installer can be used to upgrade or change settings of installed components.

5 The instance should be up and running, and you can connect to it using the MySQL workbench. See figure C.2.

Figure C.2   MySQL workbench interface

C.2   *Linux installation*

The official installation instructions for MySQL on Linux can be found at https://
dev.mysql.com/doc/refman/5.7/en/linux-installation.html.

However, certain Linux distributions give specific installation guides for it. For
example, the instructions for installing Linux on Ubuntu 14.04 can be found at
https://www.linode.com/docs/databases/mysql/how-to-install-mysql-on-ubuntu-14-04.
The following instructions are based on the official instructions.

1   First check your hostname:

```
hostname
hostname -f
```

The first command should show your short hostname, and the second should
show your fully qualified domain name (FQDN).

2   Update your system:

```
sudo apt-get update
sudo apt-get upgrade
```

3   Install MySQL:

```
Sudo apt-get install msql-server
```

During the installation process, you'll get a message to choose a password for
the MySQL root user, as shown in figure C.3.

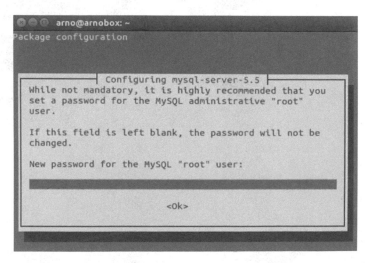

Figure C.3   Select a password for your MySQL root user.

MySQL will bind to localhost (127.0.0.1) by default.

4 Log into MySQL:

```
mysql -u root -p
```

Enter the password you chose and you should see the MySQL console shown in figure C.4.

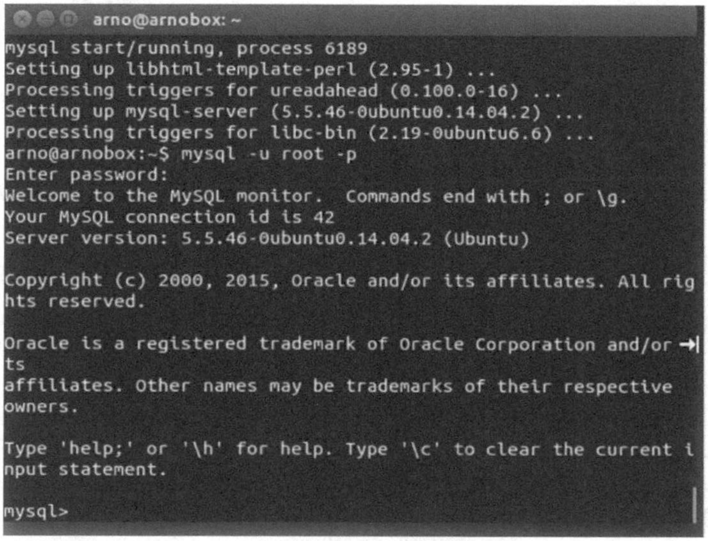

```
arno@arnobox: ~
mysql start/running, process 6189
Setting up libhtml-template-perl (2.95-1) ...
Processing triggers for ureadahead (0.100.0-16) ...
Setting up mysql-server (5.5.46-0ubuntu0.14.04.2) ...
Processing triggers for libc-bin (2.19-0ubuntu6.6) ...
arno@arnobox:~$ mysql -u root -p
Enter password:
Welcome to the MySQL monitor.  Commands end with ; or \g.
Your MySQL connection id is 42
Server version: 5.5.46-0ubuntu0.14.04.2 (Ubuntu)

Copyright (c) 2000, 2015, Oracle and/or its affiliates. All rig
hts reserved.

Oracle is a registered trademark of Oracle Corporation and/or →|
ts
affiliates. Other names may be trademarks of their respective
owners.

Type 'help;' or '\h' for help. Type '\c' to clear the current i
nput statement.

mysql>
```

Figure C.4   MySQL console on Linux

5 Finally, create a schema so you have something to refer to in the case study of chapter 4.

```
Create database test;
```

appendix D
Setting up Anaconda with a virtual environment

Anaconda is a Python code package that's especially useful for data science. The default installation will have many tools a data scientist might use. In our book we'll use the 32-bit version because it often remains more stable with many Python packages (especially the SQL ones).

While we recommend using Anaconda, this is in no way required. In this appendix, we'll cover installing and setting up Anaconda. Instructions for Linux and Windows installations are included, followed by environment setup instructions. If you know a thing or two about using Python packages, feel free to do it your own way. For instance, you could use virtualenv and pip libraries.

D.1   Linux installation

To install Anaconda on Linux:

1 Go to https://www.continuum.io/downloads and download the Linux installer for the 32-bit version of Anaconda based on Python 2.7.

2 When the download is done use the following command to install Anaconda:

```
bash Anaconda2-2.4.0-Linux-x86_64.sh
```

3 We need to get the `conda` command working in the Linux command prompt. Anaconda will ask you whether it needs to do that, so answer "yes".

D.2   Windows installation

To install Anaconda on Windows:

1 Go to https://www.continuum.io/downloads and download the Windows installer for the 32-bit version of Anaconda based on Python 2.7.
2 Run the installer.

D.3   Setting up the environment

Once the installation is done, it's time to set up an environment. An interesting schema on conda vs pip commands can be found at http://conda.pydata.org/docs/_downloads/conda-pip-virtualenv-translator.html.

1 Use the following command in your operating system command line. Replace "nameoftheenv" with the actual name you want your environment to have.

```
conda create -n nameoftheenv anaconda
```

2 Make sure you agree to proceed with the setup by typing "y" at the end of this list, as shown in figure D.1, and after awhile you should be ready to go.

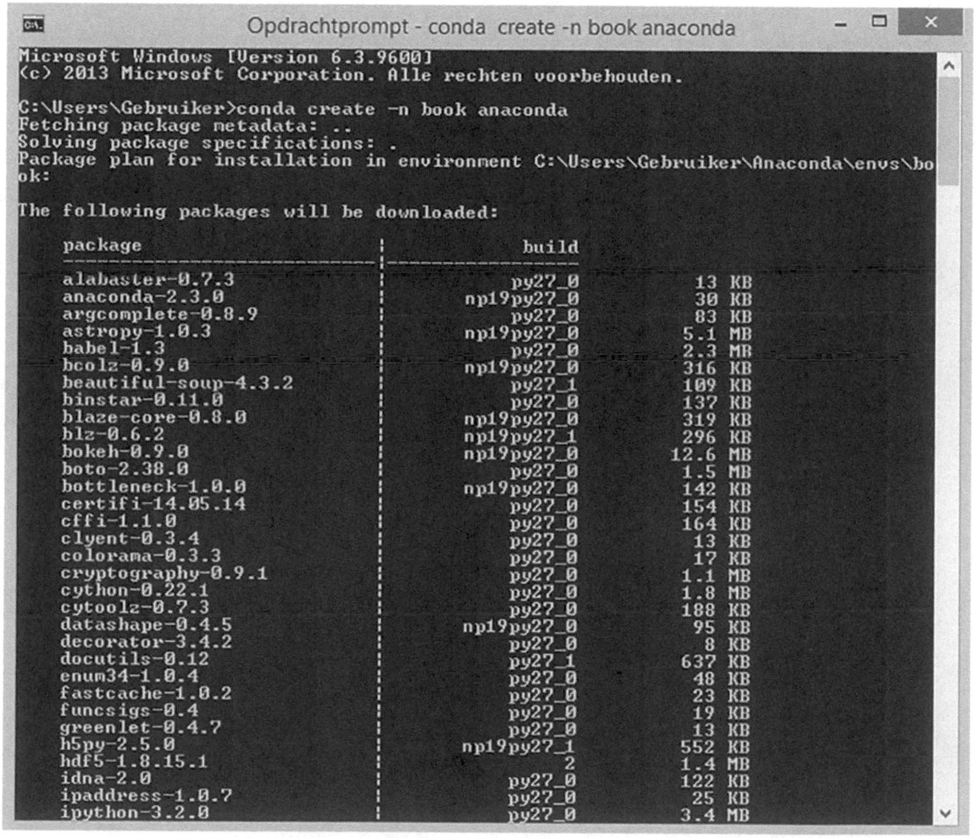

Figure D.1   Anaconda virtual environment setup in the Windows command prompt

Anaconda will create the environment on its default location, but options are available if you want to change the location.

3 Now that you have an environment, you can activate it in the command line:
   – In Windows, type `activate nameoftheenv`
   – In Linux, type `source activate nameoftheenv`

   Or you can point to it with your Python IDE (integrated development environment).

4 If you activate it in the command line you can start up the Jupiter (or IPython) IDE with the following command:

```
Ipython notebook
```

   Jupiter (formerly known as IPython) is an interactive Python development interface that runs in the browser. It's useful for adding structure to your code.

5 For every package mentioned in the book that isn't installed in the default Anaconda environment:
   a Activate your environment in the command line.
   b Either use `conda install libraryname` or `pip install libraryname` in the command line.

   For more information on the `pip` install, visit http://python-packaging-user-guide.readthedocs.org/en/latest/installing/.

   For more information on the Anaconda `conda` install, visit http://conda.pydata.org/docs/intro.html.

index